Practical Map Production

Practical Map Production

JOHN LOXTON

JOHN WILEY & SONS
Chichester · New York · Brisbane · Toronto

Reprinted with corrections July 1983

British Library Cataloguing in Publication Data:
Loxton, John
Practical map production.
1. Cartography
I. Title
526 GA105.3 80-40118

ISBN 0 471 27782 7
ISBN 0 471 27783 5 Pbk

Phototypeset by Dobbie Typesetting Service, Plymouth, Devon and printed at The Pitman Press, Bath, Avon.

Preface

This book is based on a one year lecture course entitled Map Production delivered to undergraduate students of land surveying at the University of Nairobi. It is intended not only to help students and lecturers but to be retained as a reference manual by land surveyors during those phases of their careers when they may be engaged in map production. Although based on actual or recommended practice in Kenya it is hoped that it might be useful in other anglophone developing countries (most of which are situated in the tropics).

Map producers comprise two main groups: in the first group are both official and commercial mapping agencies, whose main task is the production of new and revised basic topographical and cadastral mapping together with a limited amount of specialized mapping and mapping at small scales derived from the basic mapping. In the second group are specialized map publishers who use the national mapping as a base and edit it to suit their particular purposes. Their products include reference atlases, school atlases, motoring maps, tourist guides, and various environmental studies. Such organizations usually have to be self-financing and their staff does not normally include surveyors.

There are some differences between the two groups in training and objectives. Books on cartography have generally been written by and for the second group. This book is designed mainly for the first group although the greater part of it must be relevant to the work of all map producers.

Basic map production begins with aerial photography, ground control surveys, computing, and photogrammetric plotting. This book does not cover these stages but starts at the point where the raw material of maps (control point lists, survey field sheets, photogrammetric plots, computer tapes, etc.) is received in the cartographic office. The main stages that follow are design, drawing, printing, automation, and records.

Obviously many techniques contribute to map production, e.g. land survey, air survey, photogrammetry, computing, drawing, photography, printing, etc. On each of these and other relevant subjects many specialist textbooks and manuals have been published, which should be consulted by the reader requiring more details, always bearing in mind that there are frequent new developments in techniques, equipment, and materials; these are usually first described in professional or trade periodicals, conference reports or manufacturers' publicity literature, all of which must be obtained and studied by any map producer who wants to keep his ideas up-to-date.

In map production, the philosophy and technology are inextricably intertwined and it is both difficult and inconvenient to separate them, e.g. to describe the theory of hill shading in one part of the book and the technique of applying it in another. The various stages of map production are therefore dealt with as far as possible in the order in which they occur and both principles and practice are described at each stage.

No specimen of a completed map is included in this book. The reader is recommended to obtain a sheet of the standard 1 : 50 000 topographical map current in his own country and keep it with the book for easy reference.

Acknowledgments

The author gratefully acknowledges the kind co-operation and assistance of the following organizations and many individual members of their staffs during the drafting of this book:
Department of Surveying, University of Nairobi
Department of Surveying, University College, London
Department of Geography, University College, Swansea
Department of Cartographic Science, University of Glasgow
Department of Cartography, I.T.C., Enschede
Ordnance Survey, Southampton
Hydrographic Department, Taunton
Directorate of Military Survey, Feltham
Experimental Cartographic Unit, N.E.R.C. (now at Swindon)
Ferranti Cetec, Edinburgh
Laserscan, Cambridge
Versatec Electronics, Newbury (for permission to publish map on page 118)

Contents

List of Figures

Abbreviations used in the Text

CL	Central line of a projection
CM	Central meridian of a projection
CP	Central point of a projection
E	Easting: reduced ground distance east of zero line
IMW	International Map of the World at scale 1 : 1 000 000
JOG	Joint Operations Graphic
N	Northing: reduced ground distance north of zero line
R	radius of sphere; mean radius of Earth
TM	Transverse Mercator projection
UTM	Universal Transverse Mercator projection
c	co-latitude, polar distance
h	elevation; height above mean sea level
m_o	scale on projection radial from CP or orthogonal to CL
m_{90}	scale on projection orthogonal to direction of m_o
r	radius of a parallel of latitude in a polar projection
y	northing co-ordinate on a map projection
α	azimuth measured clockwise from north
λ	longitude
ρ	radius of curvature of a meridian
υ	radius of curvature orthogonal to a meridian
ϕ	latitude

1

Design

'Maps are among the most efficient and effective storage media devised by man; data portrayed on a medium-sized map, if presented verbally, would fill several volumes.' (Special Libraries Assn. Committee, U.S.A. 1966)

1.1 A Map: Its Nature and Purpose

A map is a picture, diagram or analogue, usually having two dimensions, of part (or all) of the surface of the Earth (or other mappable area) and is a device for transferring selected information about the mapped area to the map user. The surface of the Earth is a vast area (510 900 000 km^2) with a great variety of surface features. These may be classed either as natural topography (seas, lakes, rivers, mountains, deserts, forests, etc.) or as man-made development (towns, roads, airports, dams, plantations, etc.). None of these features is permanently fixed in size or shape: there are always small or large changes in progress. They constitute the main factor in the environment of mankind, and there is currently an increasing demand for more information about our physical environment.

Three-dimensional models can be made, either of the whole Earth (i.e. a sphere or globe) or of part of the surface (a relief model). These are often included in the definition of a map, but to avoid confusion it is best to restrict the use of the term map to a two-dimensionsal image. (Anglo-American Cataloging Rules, 1978).

Photographs taken from orbiting space satellites, or at a lower altitude from air-craft, yield much detailed information about surface features but require skilled interpretation by the user. Overlapping pairs of photos can be viewed in stereo to give a three-dimensional impression to the viewer. Photographs are used in map making; they are a valuable supplement to a map but they cannot replace it.

The detail for maps is collected both from photographs and directly from the ground area. The task of the map producer is to design the presentation of this material in map form in such a way that it can most easily be understood by the map user.

Maps are used on the ground for route planning and location of particular features, but they are also used in places remote from the map area as sources of information about the area. It is usually much quicker and easier to extract topographical information from a map in an office than to go and collect it on the ground, but for this procedure to be effective the map maker must ensure that the information shown on the map is accurate and as far as possible complete. A map has the additional merit that it can contain information that is *not visible* on the ground, e.g. place names, undefined boundaries, heights, etc.

For a general classification of types of map, see Section 1.10. For advice on map design, see Section 2.9.

1.2 The Spheroidal Earth

The surface of the solid earth, unlike a water surface, is not smoothly spread in two dimensions, but also has vertical relief such as hills and valleys. But an extensive water surface is not two-dimensional; it is curved, being part of a surface which is approximately a spheroid. Before any map can be designed, the map maker must be familiar with the qualities of this spheroid.

Analysis of the orbits of artificial satellites has yielded much information about the shape of the Earth. If the Earth did not rotate it would have nearly a true spherical shape. If the surface of the solid part were then smoothed out it would form a sphere of radius about 6368 km and if the water of the oceans were evenly spread above this, the radius of the water surface would be near 6371 km.

Rotation of the Earth about its polar axis increases the radius of the existing ocean surface at the equator to 6378 km while at the poles it decreases to 6357 km. Hence the ocean surface approximates to a spheroid (an ellipsoid of small eccentricity), but local variations in the density of crustal materials cause variations in gravity which result in small departures from a regular spheroidal shape; the resulting surface is called the *geoid*. This is the surface above which elevations are locally measured. Geodesists compute the dimensions of spheroids which best approximate to the geoid in various parts of the Earth and produce volumes of tables which the map producer should use in plotting the framework of his maps.

The circumference round the equator is 40 075 km or nearly 10 019 km for a quadrant of the equator, whereas the meridional quadrant from pole to equator is 10 002 km. This figure should be exactly 10 000 because the kilometre was originally defined as 1/10 000 of a meridional quadrant, but the measurements made two centuries ago could not achieve a better accuracy.

The difference between the polar radius and the equatorial radius is only about 1 part in 300 so that for many mapping purposes the Earth can be assumed to be a sphere. This assumption is made in the following paragraphs wherever it is necessary to keep a concept or formula simple.

Since the surface of the Earth is not plane, three dimensions are needed to fix the position of any point on it. These must be measured in a specified way from three specified zero positions. Satellite geodesy uses a set of rectangular co-ordinates with zero at the centre of the Earth (geocentric co-ordinates) but the more familiar parameters are *latitude* (ϕ), *longitude* (λ), and *elevation* (h). The first two are angles, the third is linear.

The zero position for latitude is the equator and the latitude of any point is the angle between the vertical (which is the normal to the geoid) at that point and the plane of the equator (see fig. 1.1). Latitude reaches a maximum of 90° at the poles and is usually reckoned positive northwards. Note that on a

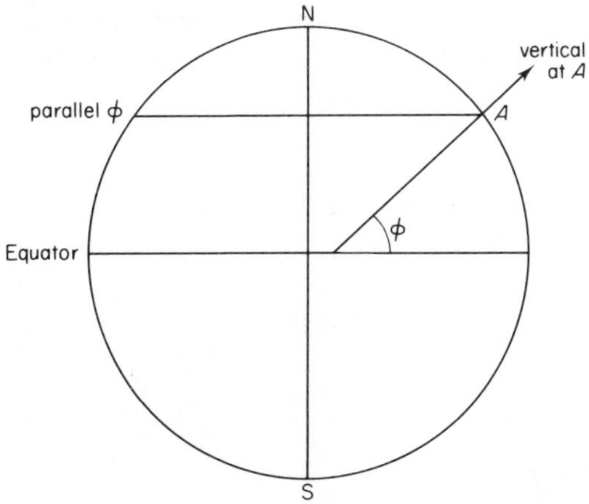

Fig. 1.1 Latitude

spheroid the vertical at any point will not pass through the centre of the spheroid (unless the point is on the equator or at a pole).

Lines of equal latitude are called *parallels*. Fig. 1.2 shows that the radius of any parallel on a spherical Earth is $R \cos \phi$ so that the length of a parallel varies from $2\pi R$ at the equator to zero at a pole.

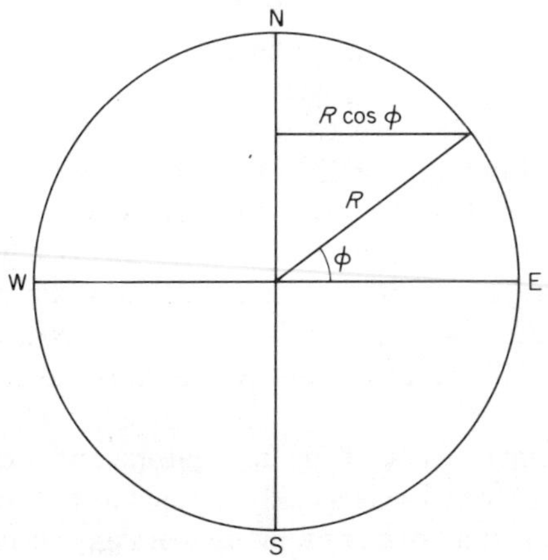

Fig. 1.2 Latitude

Any plane through the centre of a sphere cuts the surface in a *great circle* (any other intersecting plane cuts it in a *small circle*). *Meridians* are great circles passing through both poles and cutting all parallels of latitude at 90°. The difference in longitude between two points is the angle between the vertical meridian planes passing through those points. Since any meridian may be specified as zero longitude, different nations have in the past had (and some still have) different longitude systems. However for international purposes the zero used since

1884 is the meridian through Greenwich Observatory in London, England. Longitude is measured east or west from this prime meridian, reaching a maximum of 180° E or W. For computations, east is the positive direction (fig. 1.3).

The third coordinate is elevation, measured locally from a zero surface at local mean sea level. The observations have to be adjusted because the Earth is a spheroid and not a sphere.

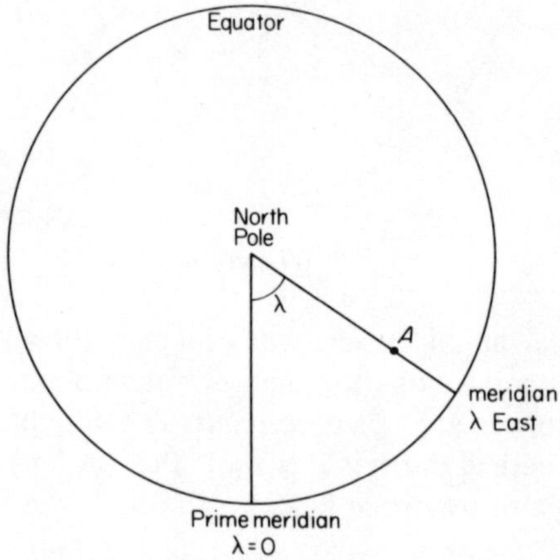

Fig. 1.3 Longitude

It is also necessary to be able to measure *direction* from a point on the surface of the Earth. In surveying this should be done as follows. The direction from point *A* to point *B* is the angle between two vertical planes at *A*, one being the plane of the meridian through *A*, the other being the vertical plane at *A* which passes through *B*. *B* can be at any altitude and may be a star. The angle between the planes is measured from zero at meridian north in a clockwise direction full circle up to 360° and is called the *azimuth* of *B* from *A*.

The term *true bearing* is used by navigators and others with the same meaning as azimuth. A bearing is usually defined as a direction measured (not necessarily clockwise) on a horizontal plane from any specified reference point (not necessarily north). The surveyor and map maker will avoid confusion if they restrict the use of the word 'bearing' to directions measured on plane maps, plans and charts (see Section 1.8.4).

Field observations of latitude and longitude are dependent on the direction of the vertical at the observation point and this may be affected by local anomalies in gravity. Hence the results are related to the geoid and must be corrected by geodesists to the spheroid.

Parallels of latitude and meridians of longitude are imaginary lines on the surface of the Earth but may be drawn as real lines on a model globe or map. The network formed by these lines is called the *graticule*.

Since latitude and longitude are angles they can be stated in centesimal measure (grades) but for international purposes (particularly sea and air navigation) the degree/minute/second system is used. The cartographer will find it useful to be able to make quick conversions of these geographical values into surface distances on the map. The most useful ones are:

at equator: 1° latitude = 110567 m, 1′ = 1843 m
at poles: 1° latitude = 111689 m, 1′ = 1862 m
at equator: 1° longitude = 111321 m, 1′ = 1855 m

Lengths of arcs of longitude at other latitudes may be obtained approximately by multiplying the lengths at equator by cosine latitude (see fig. 1.2).

One minute of latitude is a *nautical mile* which is the basic unit of distance for sea and air charts and navigation based on astronomical observations. Since it varies with latitude, the mean value 1852 m which is also the value at median latitude 45° has been adopted for international use.

1.3 Topographical Map Projections

The previous section described the shape of the Earth and noted that for many cartographic purposes it can be treated as a true sphere, using the mean radius of 6371 km. For a limited area it is more correct to use $(\varrho\upsilon)^{1/2}$ where ϱ and υ are the radii of curvature along and at right angles to the meridian in the centre of the area.

A small area of the surface of a sphere is almost plane and can be represented on a plane without distortion. On a sphere the size of the Earth, a 10 km square may be regarded as a small area. A single sheet map or plan 50 cm square could show such an area at the scale of 1 : 20000. On such a map, horizontal measurements on the ground can be plotted at their true scale lengths and directions without measurable error. (For explanation of *scale* see Section 1.3.3).

A larger area of the spherical surface cannot be truly represented in all respects on a plane. If a surveyor correctly measures the horizontal angles of a large triangle on the ground, the sum of the three angles will always be found to *exceed* 180°. On a flat map the sum of the angles of any triangle must always *equal* 180°. If the cartographer tries to preserve the angles and the area of the spherical triangle, then his map triangle would have to have curved sides.

The same sort of problem must obviously occur with other figures. Consider an area (not astride the Equator) on the sphere, bounded by two arcs of meridians and two parallels of latitude. This figure has four 90° angles but its north and south sides have unequal lengths. If it is plotted on the map as a rectangle then at least one side will be found to have a wrong length. If the four sides are plotted at their true scale lengths then at least two angles will not be 90°. It is evident that the cartographer can choose one or more qualities (shape, area, length, direction) which he wishes to preserve wholly or partly, but he cannot preserve all of them.

Many systems may be devised for transferring detail from the curved surface of a sphere to the plane surface of a map. The process is called projection because the basic methods are actual geometric projections. However, a *map projection* is defined as *any* systematic arrangement of meridians and parallels on a plane map.

Many volumes have been written on map projections and some are mentioned in the list of references. Since this book is concerned mainly with production of topographical maps (see Section 1.10) of limited areas consideration of the subject can be restricted to the fundamentals of projection theory and to details and suitability of the few projections applicable to this class of mapping. This should enable the reader to choose the projection most suited to the map to be produced; then details for the construction of the projection graticule may be obtained from one of the references and from books of tables (or computer programmes) prepared for the purpose.

1.3.1 Classes of projections

The simplest form of projection is a *perspective* projection (see fig. 1.4). A perspective centre PC is chosen from which a straight line is drawn through point A on the sphere to intersect the map plane at A'. Similarly for all other points on the sphere. If the eye is placed at PC the view of detail projected on to the plane is the same as the view from PC of the corresponding detail on the sphere. Only the distances from the eye have changed, but not the directions.

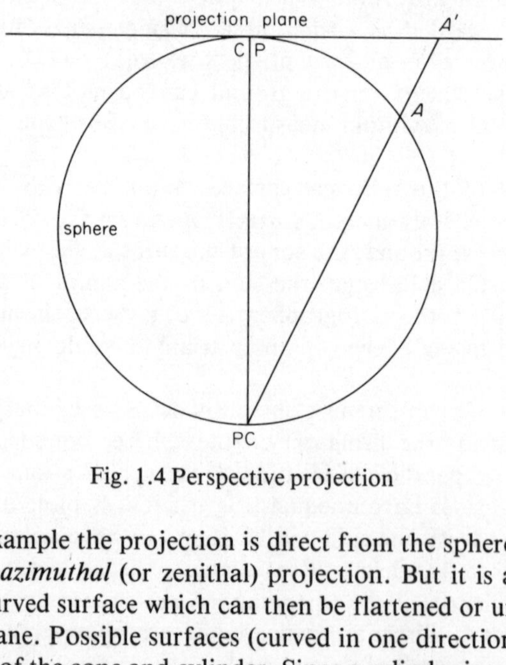

Fig. 1.4 Perspective projection

In the above example the projection is direct from the sphere on to a plane. This is called an *azimuthal* (or zenithal) projection. But it is also possible to project on to a curved surface which can then be flattened or unrolled without distortion to a plane. Possible surfaces (curved in one direction only, unlike a sphere) are those of the cone and cylinder. Since a cylinder is a cone of infinite height and a plane is a cone of zero height, it can be argued that all perspective projections are made on to a conical surface.

The cone is normally considered to be *tangent* to the sphere; it will then contact the sphere along a small circle (see fig. 1.5). If the cone flattens to a plane, the circle shrinks to a point; at the other extreme a cylinder is tangent along a great circle (fig. 1.6). The tangent point or circle becomes the *centre point* (CP) or *centre line* (CL) of the projection. Along and close to this line projected detail is the same size and shape as on the sphere.

If the tangent cone is slightly shrunk it will cut the sphere along two parallel small circles (if the cone is of zero height (a plane) the two circles coincide as one) and there will be nil distortion along both circles. This is a *secant* projection (fig. 1.7).

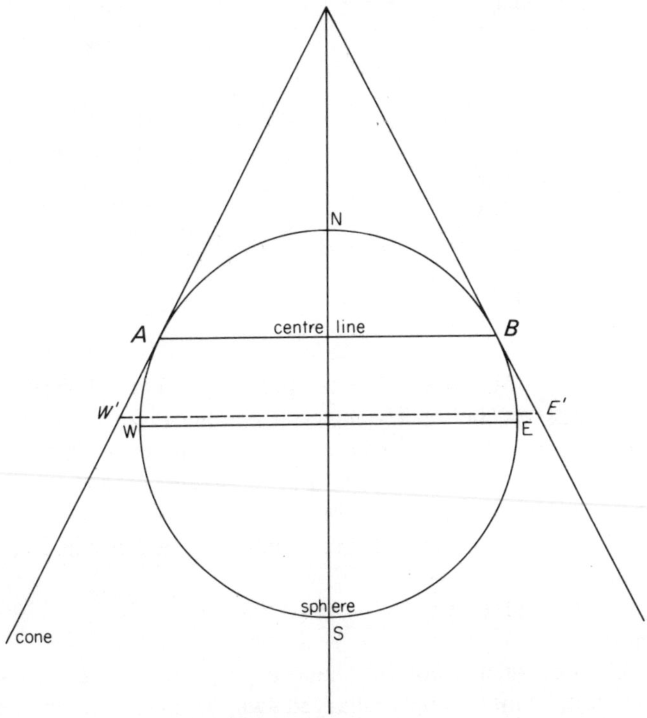

Figure 1.5 Tangent cone

The axis of the cone must always pass through the centre of the sphere and the simplest relationship between cone and sphere is when their axes coincide (as in figs. 1.5 and 1.7). This is called *normal aspect*. In projections of the northern hemisphere the apex of the cone will be at or above the north pole of the sphere and for the southern hemisphere the cone will be inverted.

If the axis of the cone (or cylinder or plane) lies in the plane of the equator the aspect is *transverse*. If the axis lies in any other direction the aspect is *oblique* (figs. 1.8 and 1.9).

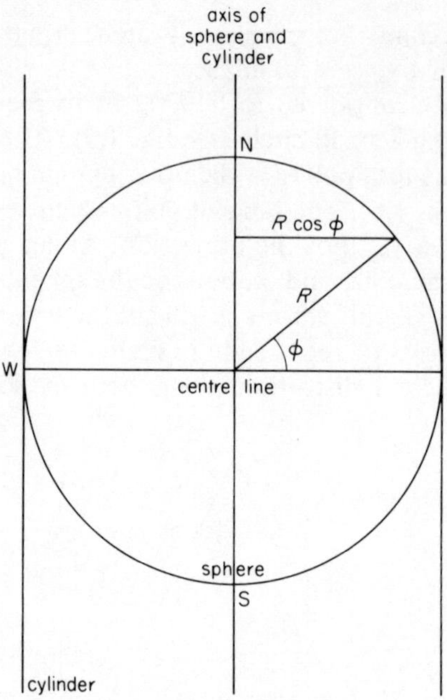

Fig. 1.6 Tangent cylinder

1.3.2 Qualities

Most projections are basically perspective and a simple perspective projection may possess the quality the cartographer requires. If it does not it can be *modified* to acquire a desired quality. There are three desirable qualities that a projection can impart to a map but in general it cannot possess more than one of them. The projection can be

(*a*) *conformal* (or orthomorphic). These come from the Latin and Greek words for true shape;

(*b*) *equivalent* (or equal-area);

(*c*) *equidistant*.

In maps of group (*a*) the correct shape of any small area is preserved but scale and area will be increasingly enlarged away from the centre line or centre point. Correct shape means that angles are preserved; in particular, meridians and parallels will intersect at 90° anywhere on the map. For topographical and general-purpose maps, this is the most desirable quality.

In group (*b*) areas are preserved but shapes are increasingly distorted away from the centre. Radial distances from CP or CL are too short while transverse distances parallel to the CL or circumferential round the CP are correspondingly too long. Equal-area maps are mainly used for presenting statistical data, e.g. population density, land use, etc.

In group (*c*) radial distances (and directions) from the CP or CL are preserved. Transverse distances and areas become increasingly too large and shapes are distorted.

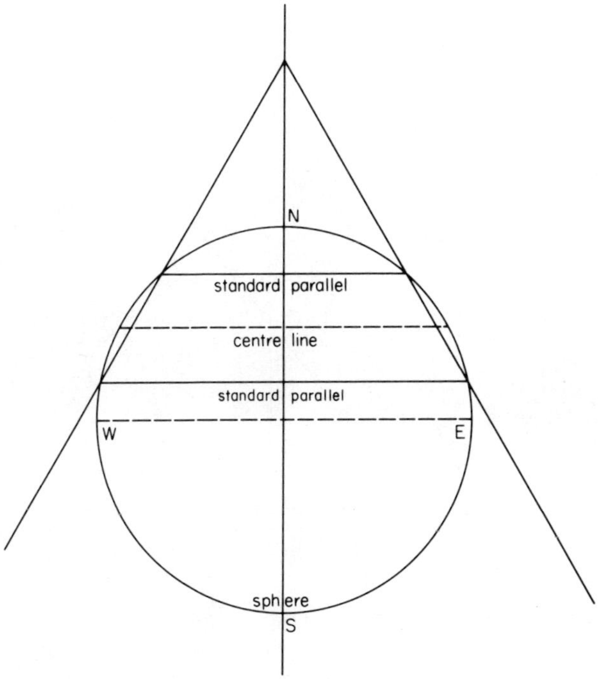

Fig. 1.7 Secant cone

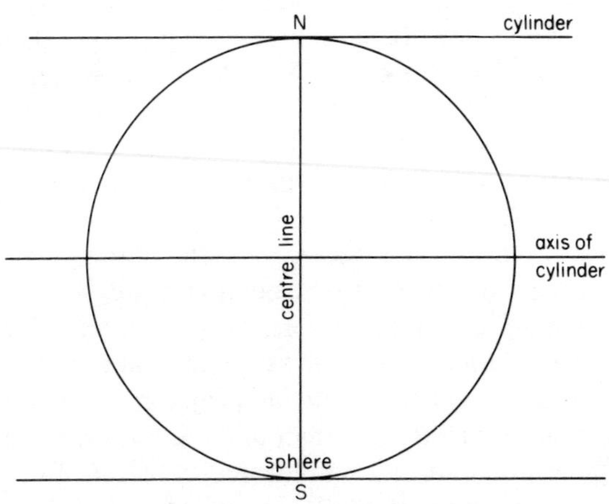

Fig. 1.8 Transverse cylinder

1.3.3 Scale, scale factor, and scale analysis

The *scale* of a map is defined as the fraction map distance/ground distance between two points. This fraction is reduced so that the numerator becomes unity; note that the *larger* the denominator, the *smaller* the scale. The previous paragraph indicates that scale is not constant all over a map. Since

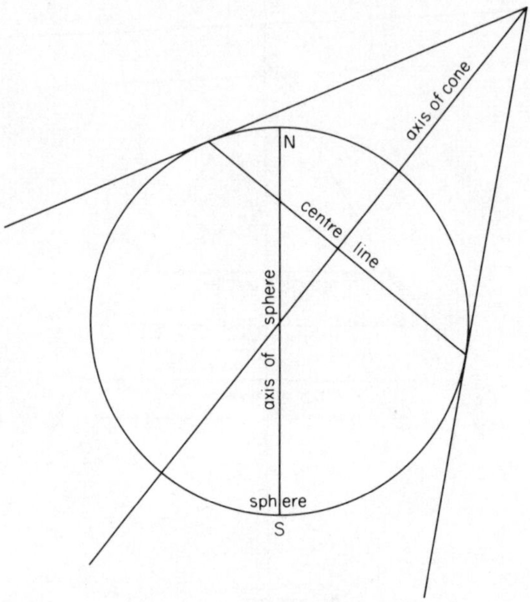

Fig. 1.9 Oblique cone

every topographical map and most other maps are labelled with their scales, the exact significance of the stated scale value needs consideration.

In a tangent projection the sphere and cone coincide along a small circle, the centre line of the projection. If the sphere is, say, 1/500 000 of the size of the Earth then the scale along the centre line of the projection will be 1:500 000. This is called the *nominal scale* or *principal* scale. In conformal projections the map scale increases away from the CL and will therefore be larger than 1 : 500000. In a secant projection the scale outside the two secant lines is also too large but between them it is too small (since the chord distance on the cone between the lines is shorter than the arc distance on the sphere). Therefore the average scale of the map is nearer to the nominal scale.

Instead of changing a tangent projection to a secant one to improve the average scale, the alternative of shrinking the tangent projection may be adopted. Either the original sphere or the projected graticule may be shrunk. The multiplier must be slightly less than unity and is called the *scale factor* of the centre line. For example, if a map covers an area 600 km square (approx. 6° of latitude and longitude) it can be shown at scale 1:1 000 000 on a single sheet 60 cm square. But if this is the scale of the centre line and the CL is in the middle of the map, a scale factor of 0.9996 must be applied there to get an average scale over the whole map of 1:1 000 000. This means that along the CL the scale becomes 400 parts in a million smaller than the nominal scale, i.e. 0.9996/1 000 000 or 1/1 000 400. The scale of exactly 1:1 000 000 will be found at about 180 km from the CL.

It may be noted that scale factor has been authoritatively defined as

$\frac{\text{actual scale}}{\text{nominal scale}}$ at any point on the projection (i.e. scale factor −1 = scale error at that point). It is arguable that variations in scale across the map should preferably be expressed in terms of *scale error* (departure from nominal scale) and the use of the term *scale factor* should be restricted to an overall change of scale as outlined above.

A scale factor of less than unity is only usefully applied to the centre line of conformal projections. To apply it to equidistant projections would defeat the purpose of the projection while to apply it to an equivalent projection would only improve the scale in one direction and worsen it in the other.

The first step in evaluating the merits of a particular projection is usually to analyse the way scale varies over the resulting map (*scale analysis*). In Section 1.3.2 above it was mentioned how scale at a point may vary according to direction of measurement. For full analysis it is necessary to be able to determine at any point the scale in any direction.

Generally, if scale at a point is not the same in all directions, the maximum and minimum values will be found in two particular directions.

(i) Radial from the centre point, or at right angles to the centre line. Scale in this direction may be designated by the symbol m_o.

(ii) At right angles to m_o; scale in this direction is designated m_{90}.

If the projection surface (cone, plane, cylinder) is in normal aspect on the sphere, m_o will be in a north-south direction, and m_{90} east-west.

If the value $m = 1$ indicates true scale, we may refer back to Section 1.3.2 and deduce that for conformal projections $m_o = m_{90}$, for equivalent projections $m_o m_{90} = 1$, and for equidistant projections $m_o = 1$.

1.3.4 The ellipse of distortion

This is a useful device which can be used as a visual indicator of scale. The supporting theory was first published in 1881 by Tissot and the ellipse is known as Tissot's Indicatrix. Tissot showed first that in any projection there must be two directions at right angles on the curved surface (sphere or spheroid) which will project as two directions at right angles on the plane. These will be directions of maximum and minimum scale at the point and are called *principal directions*. He then showed that a very small circle on the sphere will project as an ellipse on the plane and the axes of the ellipse will be the principal directions. Note that a circle is a special case of an ellipse and in a conformal projection a small circle must project as a circle since there is no distortion of small areas.

If the ellipses are drawn to a common scale at various points on the projection and have axes equal to m_o and m_{90} at each point the variation of scale and distortion across the map can be seen at a glance. (See fig. 1.10). The length of a line from the centre to the perimeter of an ellipse in any direction will be proportional to scale in that direction.

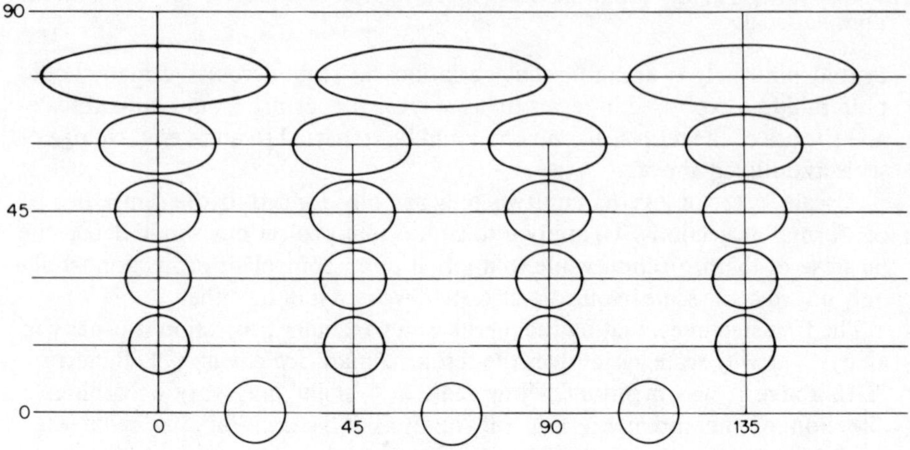

Fig. 1.10 Tissot ellipses on a normal aspect cylindrical equidistant projection

1.3.5 Angular distortion

When m_o and m_{90} at any point are known, the error in any direction from the point can easily be computed. Fig. 1.11(*a*) shows on a sphere of radius R a small triangle; A and B are two close points whose latitudes differ by $d\phi$ and their longitudes by $d\lambda$ (both in radians). The corresponding distances on the spherical surface are $R\, d\phi$ and $R \cos\phi\, d\lambda$. The angle α is the azimuth of B from A (see Section 1.2).

The spherical triangle APB projects to $A'P'B'$ (fig. 1.11(*b*)) on the plane map, the lengths of the sides AP and BP becoming dN and dE and the azimuth α becomes a plane bearing β. $\tan\alpha = \cos\phi \dfrac{d\lambda}{d\phi}$ and $\tan\beta = \dfrac{dE}{dN}$. Assume that the projection is normal aspect, then the north-south scale is $m_o = \dfrac{A'N'}{AN} = \dfrac{dN}{R\, d\phi}$

and $m_{90} = \dfrac{dE}{R\cos\phi \cdot d\lambda}$. Hence $dN = m_o R\, d\phi$ and $dE = m_{90} R \cos\phi \cdot d\lambda$.

But $\tan\beta = \dfrac{dE}{dN} = \dfrac{m_{90} R \cos\phi}{m_o R} \cdot \dfrac{d\lambda}{d\phi} = \dfrac{m_{90}}{m_o} \tan\alpha$.

This gives a relationship between direction AB on the sphere and its projection $A'B'$ on the plane. If the projection is in transverse aspect, exchange m_o and m_{90} in the formula.

We may now proceed to review briefly some of the most useful projections.

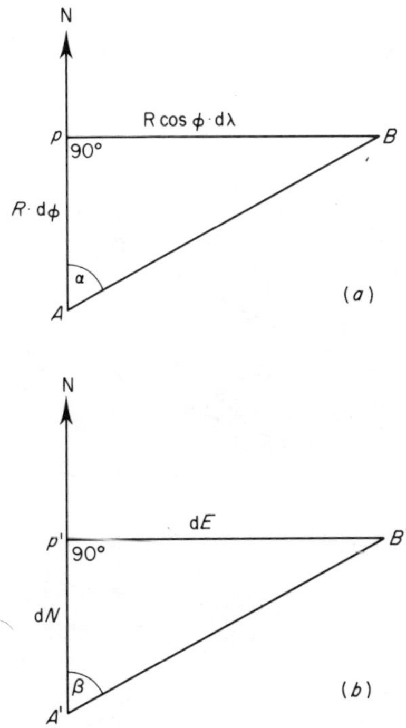

Fig. 1.11 Angular distortion

1.4 Azimuthal Projections

Azimuthal projections are most usefully applied to the polar regions, i.e. in normal aspect with the map plane tangent to the sphere at a pole. All normal aspect azimuthal projections have these common properties: all meridians are straight lines radiating from the centre point (the pole) in true directions; all parallels of latitude are circles with the pole as common centre. The only difference between the projected graticules is in the spacing between the parallels.

1.4.1 Gnomonic projection

This is a perspective projection with the perspective centre at the centre of the sphere. Projection is on to a tangent plane. Solid geometry shows that any great circle on the sphere becomes a straight line on the plane. Since a great circle is the shortest distance between two points on the surface of the Earth, this projection is valuable in navigation and is used for some planning charts of oceans. In the normal aspect with the plane tangent at a pole $m_o = \operatorname{cosec}^2\phi$ and $m_{90} = \operatorname{cosec}\phi$.

Whatever the aspect, all meridians and the equator must be straight lines; see fig. 1.12 for part of the graticule of a transverse gnomonic.

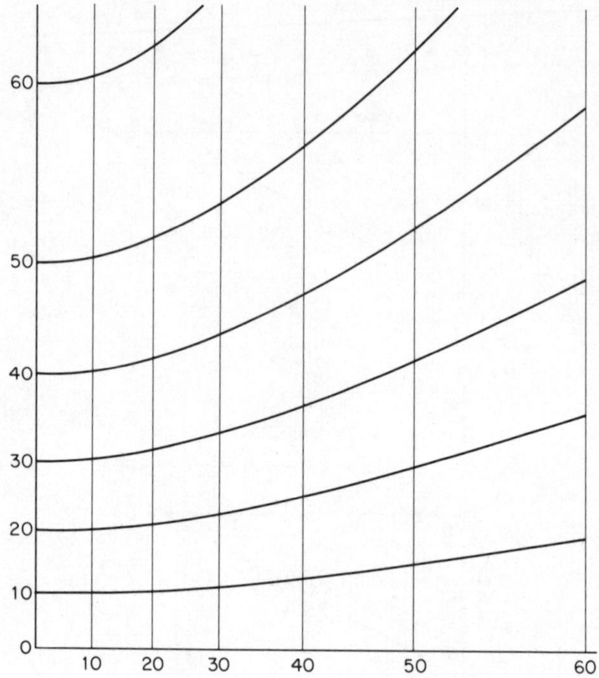

Fig. 1.12 Gnomonic projection

1.4.2 Stereographic projection

Generally only used in its normal aspect, for conformal mapping of the polar regions. The perspective centre PC is at one pole and the projection plane NA' is tangent at the other pole (see fig. 1.13). By solid geometry any circle on the sphere will project to a circle on the tangent plane so the projection is conformal. $m_o = m_{90} = \sec^2(45-\frac{1}{2}\phi)$. A scale factor may be applied to improve the average scale. It may be instructive at this point to compute m_o and m_{90} as an example of the method which may be applied to any projection.

Any infinitesimal arc of meridian AB on the sphere projects to $A'B'$ on the plane. The meridian scale m_o at A' is $\dfrac{A'B'}{AB} = \dfrac{\mathrm{d}r}{R\,\mathrm{d}c}$ where c is the co-latitude $(= 90-\phi)$ of A and r is the radial or meridian distance of A' from the central point (the pole). But $r = 2R\tan\frac{1}{2}c$, hence $\mathrm{d}r/\mathrm{d}c = R\sec^2\frac{1}{2}c$ and $m_o = \sec^2\frac{1}{2}c$. The radius of the parallel of latitude through A on the sphere is $R\sin c$ and the radius of the projected parallel through A' on the plane is $r = 2R\tan\frac{1}{2}c$. Hence the scale at A' along the parallel $= m_{90} = \dfrac{2R\tan\frac{1}{2}c}{R\sin c}$

$$= \frac{2R\tan\frac{1}{2}c}{2R\sin\frac{1}{2}c\cdot\cos\frac{1}{2}c} = \sec^2\tfrac{1}{2}c = m_o.$$

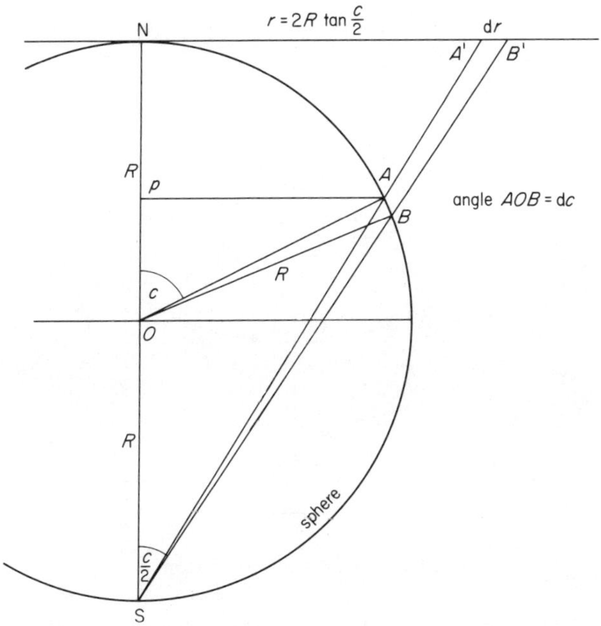

Fig. 1.13 Stereographic projection

1.4.3 Azimuthal equidistant projection

This projection is not perspective but constructed. The projection plane is tangent to the sphere at the centre point (CP). All azimuths from CP are true, i.e. all great circles through CP project as straight lines on bearings equal to the azimuths of the great circles at CP. Distances along these lines are plotted true to scale, i.e. $m_o = 1$. In the normal aspect CP is at a pole and $m_{90} = \dfrac{\frac{1}{2}\pi - \phi}{\cos\phi}$

The map can be constructed with any chosen centre point, e.g. Nairobi. Then the distance and direction from that centre to any other point on the map can be quickly found by measuring the bearing and length of a straight line from the centre to the other point. This is useful for planning aircraft flights.

1.5 Conical Projections

All projections based on a normal aspect tangent or secant cone will have these properties: all meridians are straight lines radiating from a common centre; all parallels are arcs of circles centred on the same point (see fig. 1.14). The centre line of the projection is a parallel of latitude. The pole may be a point or an arc. In a conformal projection it must obviously be a point.

Normal aspect conical projections are most suitable for middle latitudes particularly for areas having a great extent in longitude (since the scale along

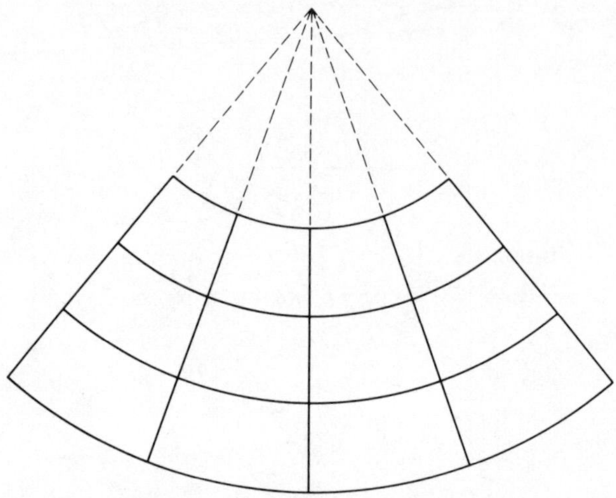

Fig. 1.14 Conical projection

any parallel is constant). Hence they are used in Russia and the USA (and for world-wide air charts constructed in the USA). Conical projections are rarely used in any other aspect but an oblique aspect may be suitable for a country whose general shape is an arc, e.g. New Zealand.

In fig. 1.5 *AB* is the centre line (tangent parallel), *EW* is some other parallel on the sphere and $E'W'$ is its projection on the cone. It is evident that $E'W'$ is always longer than *EW*, whether it is north or south of the centre line and however the projection is modified. This means that the east-west scale (m_{90}) increases away from the centre line. The only way to make the projection conformal is to increase the spacing between the parallels in such a way that the north-south scale (m_o) equals m_{90} at any point. To achieve this, the required distances of each parallel from the centre line can be obtained by a computation similar to the one for Mercator projection (see Section 1.6).

To improve the average scale, a scale factor may be applied to shrink the whole projection; alternatively the vertex of the cone may be moved down nearer to the pole to give a secant projection (fig. 1.7). Provided that the vertex angle of the cone is unchanged, the result is the same. The projection will then have two *standard parallels* along which the scale is true; between them the scale is too small and outside them it is too large. The best position for the standard parallels is such that two-thirds of the map area lies between them.

1.5.1 Lambert's projection

The German mathematician Lambert devised many projections, published in a book in 1772, but today any reference to 'the Lambert projection' may be assumed to mean the conformal conic with two standard parallels.

$$m_o = m_{90} = \frac{\tan^n \frac{1}{2}c}{\sin c} \, \frac{\sin c_1}{\tan^n \frac{1}{2}c_1}$$ gives the scale at any point whose co-latitude

is $c\ (= 90° - \phi)$ and $n = \dfrac{(\log \sin c_1 - \log \sin c_2)}{(\log \tan \frac{1}{2}\, c_1 - \log \tan \frac{1}{2}\, c_2)}$ where c_1 and c_2 are the

co-latitudes of the standard parallels. The logs are naturals to base e.

1.5.2 The polyconic and IMW

As noted above, a projection using a single cone must have increasing scale error away from the centre line. A projection published by the United States Coast and Geodetic Survey (now the U.S. National Ocean Survey) in 1855 attempted to remedy this defect by using an infinite series of cones so that every parallel is a standard parallel and the east-west scale is everywhere true. Each parallel has a different radius (= R cot ϕ) and a different centre; the centres are located by choosing a straight central meridian and dividing it truly (see fig. 1.15). All other meridians then plot as curves concave to the central meridian (fig. 1.16). The north-south scale increases away from the central meridian. The projection is in fact an equidistant one, with the central meridian as a centre line, and it is not conformal; distortion also increases away from the central meridian.

When (in 1909–1913) the project for an International Map of the World (IMW) at scale 1 : 1 000 000 was being drafted, the projection adopted was a modified polyconic. The world, between 60° north and south latitudes, was divided into sixty zones each 6° of longitude wide and each zone was then divided into map sheets each covering 4° of latitude (see fig. 1.17). The north and south edges of each sheet are standard parallels truly divided and have the same radius as in the polyconic ($r = R \cot \phi$) but they are spaced closer together. All the meridians are made straight lines from north edge to south edge of each sheet. The meridians 2° each side of the central meridian are truly divided. Between them the north–south scale is too small. In the centre of the map the east–west scale is also too small. Each sheet edge will fit the adjoining sheet edge but since all four corners of each sheet are less than 90° it is not possible to make up a block of four sheets. The projection is not conformal but errors in all qualities are kept to a minimum.

The IMW needs 1500 sheets to cover the land areas of the world. The central bureau for co-ordinating work on the map was taken over by the United Nations in 1953 and in 1962 a conference at Bonn revised the specifications, agreeing among other things that future sheets would be plotted on the Lambert projection (already adopted for the World Air Chart at 1 : 1 000 000).

The IMW has been somewhat eclipsed by the rapid production since 1964 in the Comecon countries of the 1 : 2 500 000 World Map (*Karta Mira*) and will also tend to be superseded by the progress of world cover at larger scales (e.g. 1 : 500 000 and the NATO 1 : 250 000 Joint Operations Graphic or JOG — see Section 3.3). The IMW sheet lines are unsuitable for many countries, e.g.

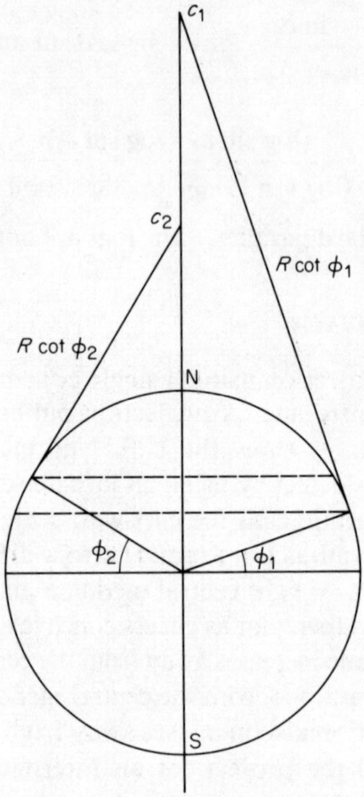

Fig. 1.15 Polyconic projection

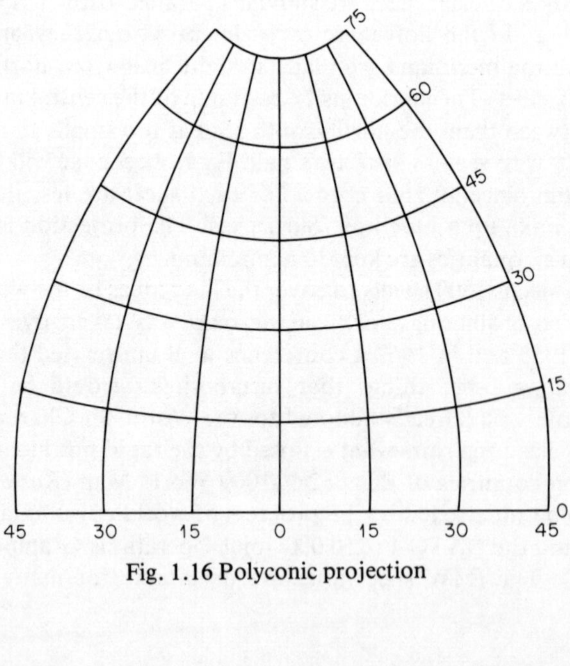

Fig. 1.16 Polyconic projection

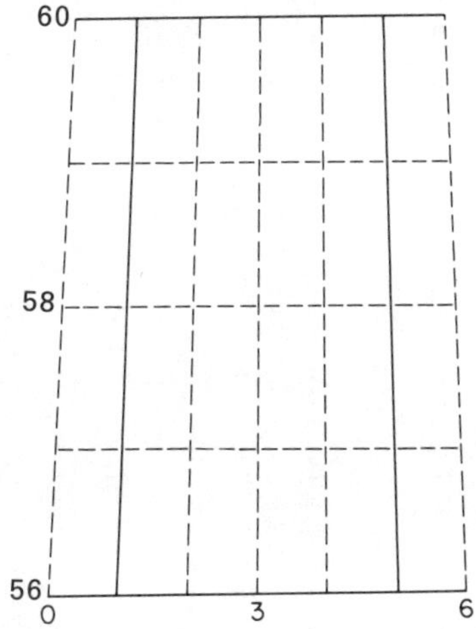

Fig. 1.17 A sheet of the IMW. Solid lines are true to scale

Kenya falls in parts of seven IMW sheets but can be (and is) covered by only two slightly larger 1 : 1 000 000 sheets with rearranged sheet lines.

The Karta Mira uses in each hemisphere one equidistant secant conic for latitude zones 0–24°, another one for 24–72°, and an azimuthal equidistant for the polar region. The sheets are large, each covering nine IMW sheets in latitudes 0–48° and twelve IMW sheets in higher latitudes. The full series is 244 sheets covering both land and ocean.

1.6 Cylindrical Projections

All projections on to a normal aspect cylinder have the following properties: all meridians are straight parallel north-south lines with equal spacing between them; they intersect at right angles the parallels of latitude which are all straight east–west lines. All the parallels are equal in length to the equator and therefore the east–west scale must increase away from the equator. For a tangent cylinder the equator is the centre line and the scale along any other parallel of latitude ϕ is sec ϕ. The spacing between parallels of latitude and hence the scale along the meridians depends on how the projection is modified to give a desired quality (see fig. 1.6).

1.6.1 Mercator's projection

If the projection on to a normal aspect tangent cylinder is made conformal, the projection is known as the Mercator projection. It was already in use in the sixteenth century (1569) and is probably still the most familiar and most used map of the world. It has a uniquely valuable property: a line of constant azimuth on the sphere projects on the map as a straight line (a loxodrome or rhumb line or line of constant bearing) cutting all meridians at the same angle.

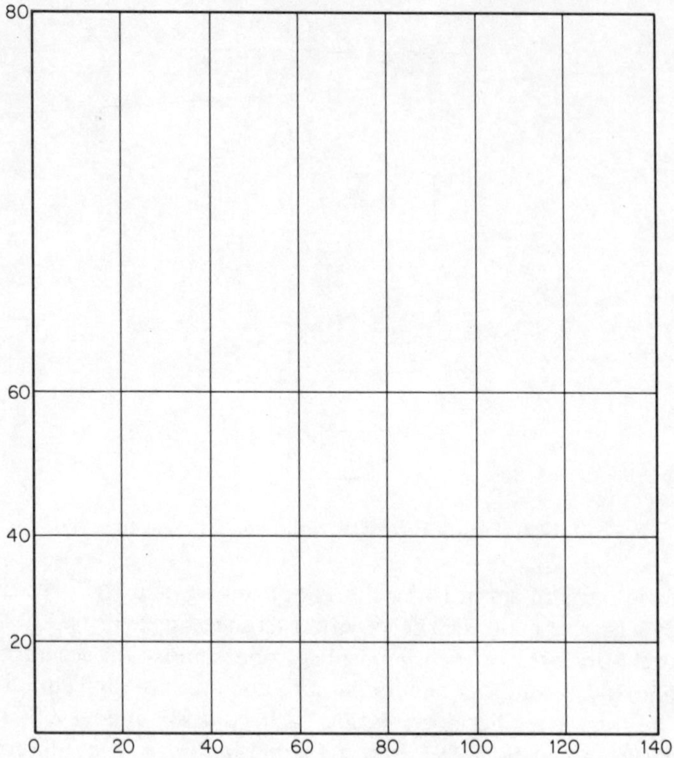

Fig. 1.18 Mercator projection graticule

This property makes it still the most used projection for navigation charts (see fig. 1.18).

There is no perspective projection on to a cylinder which is conformal and it is necessary to compute the spacing from the equator to the parallels.

Consider an infinitesimal arc of meridian at latitude ϕ. The north–south scale is length on map/length on sphere $= \dfrac{dy}{R\,d\phi}$. Similarly the east–west scale is $\sec\phi$ and for conformality these must be equal. Hence $dy = R \sec\phi\, d\phi$. Integration gives $y = R \log_e \tan\left(\dfrac{\pi}{4} + \tfrac{1}{2}\phi\right)$, y being zero at the equator. This expression gives the distance on the map from equator to parallel of latitude ϕ for a Mercator projection of a sphere of radius R.

The projection is suitable for the topographical mapping of a country lying across the equator (e.g. Kenya) but in fact is rarely so used.

1.6.2 Scale error in conformal projections

If N is the distance from the equator along a meridian arc to latitude ϕ, then

ϕ (radians) may be written N/R. The scale at latitude ϕ on the Mercator map is sec ϕ, or sec (N/R). This may be expanded as the series $1 + \frac{N^2}{2R^2} + \ldots$ For values of N that are small compared with R we may ignore higher power terms in the series. Then the term $\frac{N^2}{2R^2}$ represents the scale error at distance N from the equator. This shows that the error is proportional to the square of the distance from the centre line. This statement is true of all the conformal projections, i.e. stereographic, Lambert, Mercator whether in normal, transverse or oblique aspect.

1.6.3 Transverse cylindrical projections

Whereas the azimuthal and conical projections are rarely used in other than normal aspect, the transverse cylinder produces what is nowadays regarded as the most important projection for medium-scale topographic mapping, possibly even for all mapping at scale 1 : 1 000 000 or larger.

A transverse cylinder is tangent to the sphere along a meridian which becomes the centre line on the projection. It is logical to choose a meridian which bisects the area to be mapped. In Britain the original large scale mapping was done by groups of counties and each group had its own central meridian. When resurvey on a national basis was started, meridian 2° W was chosen as central meridian (CM) for the whole country.

The central meridian projects as a straight line, other meridians as curves concave to the central meridian, cutting the equator at right angles and converging to the poles. The equator projects as a straight line; other parallels as curves convex to the equator and cutting the central meridian at right angles (see fig. 1.19).

1.6.4 Cassini projection

The simplest modification on the transverse cylinder is the equidistant mode (see fig. 1.20). A point A on the ground (i.e. on the sphere) is distant E from the central meridian OP. Angle APO is 90°. P is distant N from the equator at O. On the projection O$'P'$ and $P'A'$ are plotted scale-equal to OP and PA. This looks very simple and satisfactory; the equator and the central meridian are zero lines for a system of rectangular co-ordinates and other graticule lines can be ignored. The projection was first proposed by Cassini in 1745 and widely used for two centuries. However, on the sphere, PA is a great circle which is converging towards the equator (just as meridians converge from the equator to a pole) while $P'A'$ is parallel to the map equator. Hence A' is too far from the equator and the N–S scale at A' is too great. As noted in Section 1.6.2 above, the error increases with the square of the distance from the central

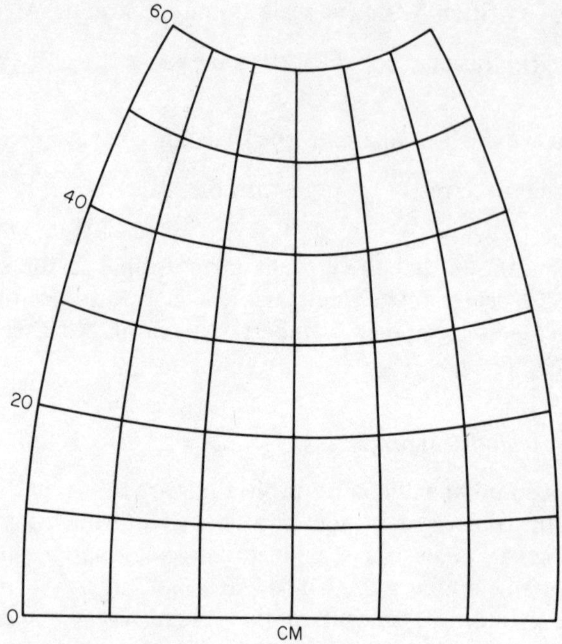

Fig. 1.19 A transverse cylindrical projection graticule

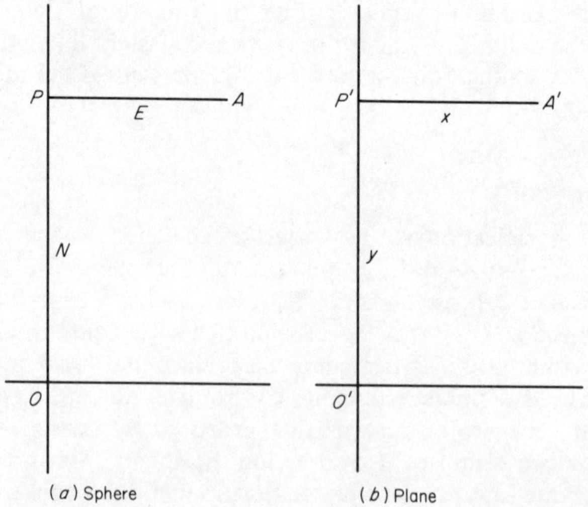

Figure 1.20 Cassini projection

meridian. At 200 km it reaches 1 in 2000 and mapping should not be plotted at greater distances.

1.6.5 Transverse Mercator

To make the projection conformal, the east-west scale must be increased to equal the north–south scale at any point. By analogy with normal aspect Mercator, scale at A' is $\sec\frac{E}{R} = 1 + \frac{E^2}{2R^2} + \ldots$ and the distance $P'A' = R \log_e$

$\tan \frac{1}{2} \left(\frac{\pi}{2} + \frac{E}{R}\right)$. This modification was developed by Gauss and published by

Lambert in 1772. It probably should be called the *Gauss Conformal* but is more commonly known as Transverse Mercator (TM). As with other conformal projections the average scalc can be improved by projecting on to a secant cylinder or applying a scale factor. If the map is shrunk by 1 in 2000 it is possible to map a zone 600 km wide with an error of –1/2000 at the centre and + 1/2000 at the edges, correct nominal scale being found 200 km each side of the centre line.

1.6.6 Universal Transverse Mercator

Surveyors and cartographers were slow to appreciate the merits of TM and it gradually came into use on a local basis, each survey or map production organization choosing its own central meridians, zone widths, and scale factors.

However, the present world tendency to international standardization has not failed to make itself felt in the field of map projections. It was proposed in about 1950 and speedily adopted by many nations that a Universal Transverse Mercator (UTM) projection might be used as a basis for world-wide topographical mapping. Among other advantages is that one book of tables will suffice for plotting the graticule of many map sheets (provided that they are in areas using the same dimensions for the figure of the Earth).

The world (except the polar regions) is divided into sixty zones each 6° of longitude wide. Starting from 180° W the zones are numbered eastwards 1 to 60. Central meridians are at odd multiples of 3°. A standard scale factor of 0.9996 is applied which makes the scale true at 180 km each side of each central meridian. Obviously the zones are wider in the tropics and the error at the edge of a zone is about 1 in 1500; at latitudes greater than 40° the scale error does not exceed 1 in 2500.

It may be noted that the UTM zones coincide with IMW sheet lines. UTM could perhaps replace the projection of the IMW although the meridian sheet lines are slightly curved so that sheets adjoining east–west would not exactly fit.

1.7 Choice of Projection

In planning a new map, or series of maps, of a limited area (e.g. a national territory) one of the first decisions to be made is choice of a projection. It is likely that there are existing maps at other scales of the same area and hence a good reason for using a projection already in use. However this may be a good opportunity to review the merits of that projection and to start changing to another one.

UTM is currently being adopted in preference to other projections except where it is obviously unsuitable for an area. It is conformal, can be used at all inhabited latitudes and there are tables based on at least five different spheroids for plotting graticules and grid. It can also be used for statistical maps because the area errors are small enough to be acceptable.

Most maps at scales smaller than 1 : 1 000 000 extend over more than one UTM zone and so also may maps at larger scales if the area mapped crosses a zone boundary. In this case, if the extent of the area in latitude is greater than in longitude, a Transverse Mercator with central meridian bisecting the area may be the most suitable.

If the extent in longitude exceeds latitude then the Lambert conformal with two standard parallels is more suitable. It may be noted that this projection can be used in equatorial countries, with one standard parallel each side of the equator (but not at equal latitudes as this could give a secant cylindrical projection).

Although equal-area projections are preferred for statistical mapping, the area error in a conformal projection is unlikely to become significant within the limited areas under consideration.

1.8 Map Grids

On the curved surface of the Earth, position is defined by the universal system of latitude and longitude. All international, and most national, geographical gazetteers use this system for locating listed places. It is normal practice for small-scale maps (1 : 500 000 and smaller) to carry a graticule of meridians and parallels to assist in locating positions of places from their geographical coordinates and vice versa.

However, graticule lines are usually curves and/or the spacing between them varies and map measurements between them made with a common scale (graduated probably in mm) have to be converted to angular measure. They are not a suitable reference system for rapid and precise position fixing which is often required on medium- and large-scale maps.

The most efficient reference system on a plane is the well-known Cartesian coordinate system based on two sets of parallel grid lines, the directions of the two sets being at right angles. In mathematical usage one set runs in an x direction and one in a y direction but to avoid ambiguity in mapping it is better to adopt east and north as the prime directions (skew grids running in other directions should be avoided). One line in each direction is given value zero and lines parallel to it are numbered, increasing positively to the east and north and negatively west and south.

Arbitrary grids are possible and are often used on tourist maps and other commercial mapping, but for optimum value there must be a known and simple relationship between graticule and grid to facilitate conversion of geographical to grid coordinates and vice versa. If a pair of graticule lines can be found which intersect at right angles and are both straight, there are obvious advantages in choosing them as zero lines for the grid.

On transverse azimuthal and transverse cylindrical projections there is no choice: only the central meridian and the equator are straight.

On normal cylindrical projections all graticule lines are straight so that the equator and any one meridian can be chosen.

On normal aspect azimuthal (polar) projections all meridians are straight; the usual zero axes are the 0–180 and the 90–270 meridians. Positive directions are optional.

In conical projections there are no straight parallels; the grid zero should be on the central parallel, one axis being the meridian through that point, the other axis being tangent to the parallel at that point (see fig. 1.21).

1.8.1 False origin

Whatever projection is chosen for mapping a given area, the centre line of the projection will tend to bisect the area. This centre line is also likely to be a grid zero line, so that all the coordinates on one side of it will be negative. This causes unnecessary inconvenience in computing and other operations. It is therefore usual to remove the zero of grid numbering to a point beyond the SW corner of the area mapped, whence all north and east coordinates in the area become positive. The new zero point is called the *false origin* (see fig. 1.21).

1.8.2 UTM grid

Refer to Section 1.6.6. The zero grid lines for any UTM zone are the central meridian and the equator. The false origin is 500 km west of the central meridian; this is about 4½° longitude west of the CM at the equator and more at other latitudes. For the southern hemisphere the false origin is 10 000 km south of the equator (which puts the false zero line near the South Pole); for the northern hemisphere the equator remains the zero line (see fig. 1.22).

1.8.3 Grid spacing

For easy reference, measurement, or estimation by eye of distance from grid lines to detail, grid spacing must be fairly close; but too dense a grid will obscure map detail. Current practice on topographical maps tends to use grid squares varying from 1 cm to 5 cm; 1 cm = 1 km at 1 : 100 000, 2 cm = 1 km at

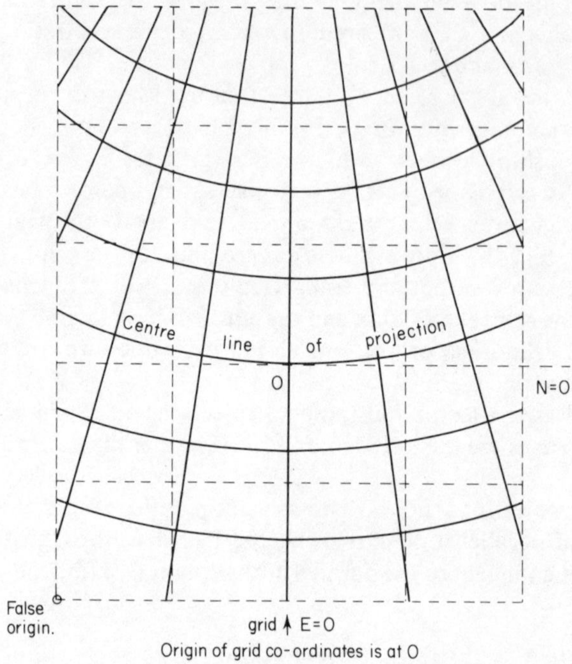

Fig. 1.21 Grid and graticule on a conical projection

1 : 50 000, 4 cm = 10 km at 1 : 250 000, etc. Every tenth grid line should be thickened for ease of use. See Section 4.4 for grid plotting.

Although in the past various units have been used for grids, it is both desirable and probable that all future grids will be numbered in metres.

Grid line values are usually placed (as also are graticule values) in the margins of the map, but for rapid reading it is convenient to have the values printed on the face of the map at 10 or 20 cm intervals. For this purpose, only significant figures need be shown, surplus zeros being omitted.

1.8.4 Direction on a map

It was noted in Section 1.2 that direction on the sphere is measured by *azimuths*. On any projection other than Mercator the graticule is not a suitable datum for measuring direction. Measurement is therefore made from grid north which is a constant direction all over a map. The direction of point B' from point A' is the angle at A' between grid north at A' and the straight line on the map $A'B'$. This angle is called the *grid bearing* of B' from A' and is measured clockwise from grid north. *The reverse bearing* $B'A'$ differs from the forward bearing $A'B'$ by 180° (unlike a reverse azimuth). The angle between grid north and a meridian at any point on the map is called the meridian *convergence* at that point.

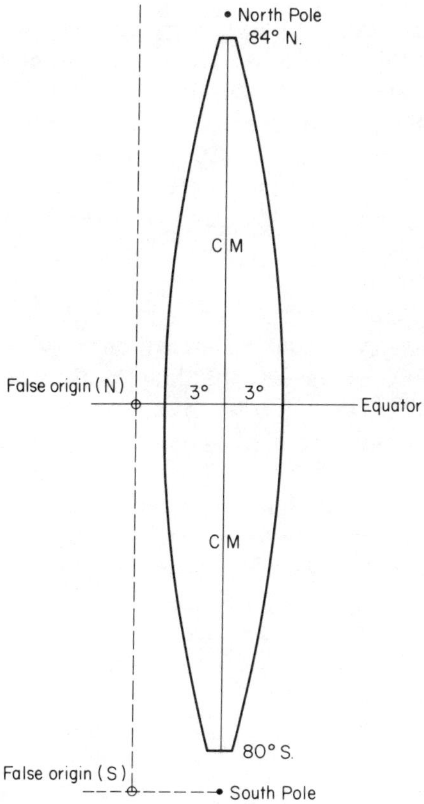

Fig. 1.22 A UTM grid zone

1.9 Scales

1.9.1 Scale indicators

The scale of a map is the ratio or fraction of paper distance/ground distance and can be variously stated in the forms:

$$\frac{1}{5000} \qquad 1/5000 \qquad 1:5000$$

The first is cumbersome. In manuscript the second is the quickest to write. In typing the second and third have equal merit. The generally preferred form is the third.

With the widespread adoption of metric measurement and standard round-number scales, old style scale statements like 'one inch to one mile' are (or should be) passing out of use. See also Section 2.7.8.

1.9.2 Standard scales

Given the dimensions of an area to be mapped and the size of paper on which the map is to be printed, an approximate scale is given by the ratio between the longest dimension of each. This can then be rounded off to some convenient figure. In this way maps have been produced at a great variety of scales showing little systematic relationship.

But a complete national map coverage will include maps and map series at a wide range of scales from perhaps 1 : 2500 for town maps to 1 : 50 000 in rural areas and 1 : 1 000 000 to cover the whole country in a few sheets. In map production there is a frequent need to transfer detail from large-scale maps to smaller scales so that there is obvious advantage in having a simple relationship between the scales.

It has been internationally recommended (in the SI system of standards) that a set of standard scales should be based on powers of ten: 1 : 10 000, 1 : 100 000, 1 : 1 000 000. These can be doubled to give 1 : 5000, 1 : 50 000, 1 : 500 000 and doubled again: 1 : 2500, 1 : 25 000, 1 : 250 000. Instead of the second doubling the original set may be halved to give 1 : 20 000, 1 : 200 000, 1 : 2 000 000.

There is no international agreement on the divisions between small, medium, and large scales but the following tabulation is a rough guide and to some extend correlates with the types of map for which the scales are most used:

Scale:	1 large	2 medium	3 small
(a)	1 : 2500	1 : 25 000	1 : 250 000
(b)	1 : 5000	1 : 50 000	1 : 500 000
(c)	1 : 10 000	1 : 100 000	1 : 1 000 000
(d)	1 : 20 000	1 : 200 000	1 : 2 000 000

1.9.3 Choice of scale

Scales 1 (a) and (b) are used for basic mapping of built-up areas (even larger scales may be used in densely developed spots). Scales 1(c) and (d) may be used for maps derived from (a) and (b) to show a whole town on one map sheet.

Scales in column 2 may be used for basic mapping of rural areas, ranging from (a) for areas of dense development to (d) for deserts. Scales in column 3 are for maps of large areas derived from column 2 maps and are all used for various world series (as is the even smaller scale of 1 : 2 500 000 used for the Karta Mira — see Section 1.5.2).

1.10 Classes of Maps

Reference has been made in previous paragraphs to basic maps, derived maps, statistical maps, navigation charts, etc. These terms should now be defined more fully.

Topographical maps show the visible surface features of land areas or as many of them as the scale will permit plus some invisible detail such as names, descriptions, boundaries, etc. Topographical maps not only show planimetric position but also indicate relief by some method.

Basic mapping is topographic mapping made from new surveys and is usually at a larger scale than any previous mapping of the area.

Derived mapping is topographical mapping made at smaller scales from basic mapping and thus each sheet covers a larger area than a basic map sheet.

Special mapping is a rather nebulous term that covers almost every sort of map other than general topographic maps.

Thematic (or topical) maps illustrate a particular theme or topic such as rainfall, land use, population density, etc. The statistical information on such themes has to be obtained from sources outside the map production organization.

Cadastral maps are a class of thematic map though they are often classed separately. They show property boundaries and areas with identification numbers and have legal significance in land registration, land taxation, etc.

Charts are a map form adapted specially for navigation. Nautical charts cover water areas not covered by topographical maps and show much invisible detail such as depths, currents, etc. (see Chapter 3).

Plans cover small areas at a large scale. They do not need to be plotted on a projection. They usually do not give relief information.

1.11 Map Size and Shape

Maps appear in many sizes and shapes ranging from small pocket atlases to large wall maps. But for national series topographic mapping it is desirable to have some standardization. There is now an international series of paper sizes starting from size A0 with an area of one square metre (see fig. 1.23). A0 is halved to give A1 which is halved to give A2 and so on. All the sizes in this series have the same shape, which requires that the sides must be in the ratio $1 : 2^{1/2}$ (1 : 1.414 . . .). Hence the sizes (in mm) are:

A0	1189 × 841
A1	594 × 841
A2	594 × 420
A3	297 × 420

For economy in production, a map sheet should be as large as possible. Two factors put an upper limit on size: the capacity of available printing presses, and ease of handling. In the A series it is found that A1 is the largest which is easy to handle. Many charts are larger than this but most maps are smaller — 60 × 84 cm may therefore be accepted as a suitable standard paper size. Note that 'the size of the map is given in cm' (Anglo-American Cataloging Rules 1967).

For a single-sheet map the direction of the greatest length of the paper is determined by the shape of the area mapped but for series maps there is a choice. In practice, maps are easier to handle and to read when the greatest length is horizontal or left–right or east–west.

In considering paper size it must be realized that the area of mapped detail does not fill the whole sheet. The printing machine cannot print right to the edge of the paper. It is possible to trim off any blank area after printing but this is only done if there are compelling reasons for it. In addition, space

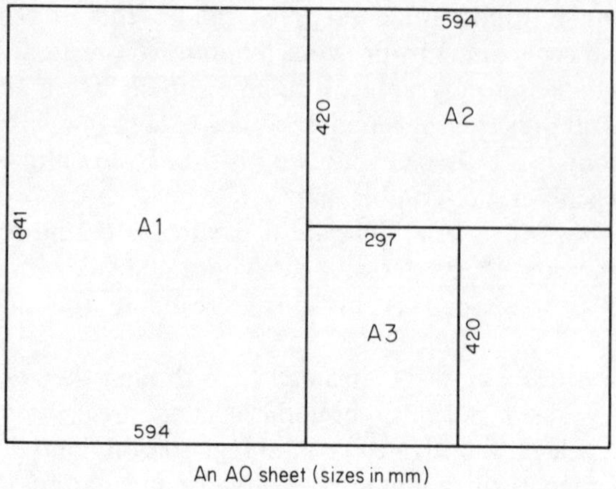

Fig. 1.23 International paper sizes ('A' series)

is required outside the map-detail area for printing other information.

1.11.1 Map series sheets

Basic mapping of a large area normally requires many sheets, all at the same scale and in the same style, and nearly all of the same size and shape. Generally they must all fit together without gap or overlap; it follows that the shape of a sheet must be approximately square or rectangular with greatest length east-west. As an example, consider a series at scale 1 : 50 000 which is the commonest scale for rural basic mapping in the developing countries.

Sheet edges are normally either graticule lines or grid lines. The latter give sheet edges which are truly square or rectangular but if the mapping area covers more than one grid zone there are problems with sheets on the zone boundary. Hence in low latitudes the chosen edges are usually graticule lines.

Within the tropics, 110 km can be taken as a round figure for the length of a degree, equal to 220 cm on a map at 1 : 50 000 scale, or 55 × 55 cm for a quarter degree square. This fits into the standard 60 × 84 cm paper size leaving sufficient space for a margin and marginal information. Although somewhat different paper sizes are widely used, the quarter degree square is the most common unit for 1 : 50 000 mapping in low latitudes. A half degree square at 1 : 100 000 or a degree square at 1 : 200 000 also give this size of map.

Considering now the scales in column 3 of Section 1.9.2, there is already an internationally agreed shape and sheet lines for the 1 : 1 000 000 map (see Section 1.5.2). The 4° × 6° format at the equator measures 442 × 668 km giving a map area 44 × 67 cm. This can also fit comfortably into A1 paper. The same map area is obtained for the other three scales in column 3 by suitable change in the dimensions of the graticule, i.e. to 1 × 1½, 2 × 3, or 8 × 12 degrees respectively.

This format and map size could of course be used for the medium scales instead of the square format, but the sheet lines would have inconvenient values: a 1 : 50 000 sheet would be 12 × 18 minutes or arc, and 18′ is not a submultiple of 1°. Whatever the dimensions adopted, any one sheet should cover the same area as an exact number of sheets at the next larger scale.

When the mapped area has a square format, the long axis of the rectangular paper may run either way. Quite often it is found to run north-south, with the information panel placed below the map. The only argument in favour of this arrangement is that it looks symmetrical. The long axis should run E–W with the information panel on the left. If a map series is placed in a file or folder, the binding margin is on the left, and the map area of each sheet is more accessible being on the right-hand, or opening, side of the folder. If the series sheet edges are to be metric grid lines, the most economical format for printing on A1 paper size is 10 × 20 km at scale 1 : 25 000 and pro rata for most other scales (e.g. 20 × 40 km at 1 : 50 000). This gives a map area 40 × 80 cm in a 60 × 84 cm sheet. The long axes of both paper and map should be east–west with the reference legend printed below the map. At some scales a square 50 × 50 cm map area map be preferable: e.g. at 1 : 20 000 this covers a ground area 10 × 10 km instead of 8 × 16 km.

1.11.2 Map orientation

It is a firm convention to print the words and figures on a map so that it reads with north at the top of the sheet. There is probably a connection between this and the fact that the magnetic compass points in a northerly direction, but in the Dark Ages, when it was thought that the Earth was flat, the direction of sunrise (oriens) was chosen as the prime direction and maps were drawn with east at the top. Hence latitude (from the Latin word for width) varied across the width of the map and longitude referred to the length of the map from top to bottom. Placing the map with east at the top was called orientation, which word is still used for getting directions correct.

However, there are occasions when the convention may be ignored. If an area to be mapped on a single sheet has its greatest length in a direction other than north–south or east–west then it is economical to have the greatest length of the paper running in the same direction. Examples are the 1 : 5000 map of Entebbe (NE at the top) and the 1 : 14 000 map of Mombasa Island (NW at the top).

1.12 Summary

The preceding paragraphs have examined most of the factors which must be considered before the construction and drawing of a map, either single sheet or series, can commence. At this point decisions will have been made on projection, grid, scale, paper size, sheet lines, and general layout.

The above items constitute part of the *specification* for a map. The remainder of the specification will cover the presentation of the map detail by listing such things as colours, line widths, symbol designs, lettering styles, etc. Therefore we must now proceed to examine carefully the contents of a map.

2

Drafting Detail

2.1 The Language of Maps

Ability to transmit and receive ideas expressed in written words is literacy; similar comprehension of figures and mathematics generally is numeracy; the ability to put information into, and extract it from, maps and diagrams may be termed graphicacy. In all three media the ability of the reader to extract full and accurate information is largely dependent on the skill and clarity of expression of the author.

The language of large-scale maps (say 1 : 2500) is fairly simple. Most of the detail surveyed in two dimensions can be drawn at its true plan size and shape and there is enough space among the detail for adequate descriptions of buildings, etc., to be written in full, e.g. Primary School, Mental Hospital.

It is when scale is reduced that cartography has to develop its own special language and the skill of the cartographer is most exercised. As much useful information as possible has to be shown in an ever-decreasing space and it must still be clearly legible and comprehensible. This requires several forms of processing: selection, generalization, exaggeration, and symbolization.

2.1.1 Selection

Selection means choosing which items of detail to omit and which to retain as scale decreases. The criterion may be actual physical size, or it may be importance. The choice will be affected by the purpose of the map, by editorial policy, or even by personal idiosyncrasy.

2.1.2 Generalization

Generalization means both simplification and combination. For example, small bends in a river disappear and are replaced by a smoother line. Several adjacent buildings are merged and shown as a continuous built-up area.

2.1.3 Exaggeration

This is best explained by an example. Consider a main road 10 m wide, selected to appear on a map at scale 1 : 100 000. If truly plotted the width of the line representing the road on the map would be 0.1 mm wide. This is not only difficult to draw, print and read but also such a fine line would lack the prominence among other map detail which its importance merits. The road will be shown by a line at least 0.5 mm wide, i.e. a minimum exaggeration of five times.

Where detail is crowded, exaggeration can cause overlapping, e.g. a road running alongside a railway alongside a river. One solution is to omit the least important feature; if this is not practicable, then one or more features must be displaced. Point features are usually the first to be sacrificed (except survey control points which must remain accurately positioned). The ascending order of priority for some other detail is usually vegetation, buildings, water, roads, railways.

2.2 Symbolization

In the language of maps the equivalent of shorthand is symbolization. Its main purpose is to replace written descriptions. Instead of drawing a plain black line on the map and annotating it as 'standard gauge double-track railway' a line of a particular width and pattern is drawn, and the meaning of that particular style is explained in the margin of the map.

The science of communication by signs and symbols is called semeiology and various aspects of the subject are debated frequently and at length in cartographic professional periodicals and conference papers. Most of the discussion is applicable to thematic maps rather than to topographical maps. There is an international agreement on the symbols to be used on sheets of the IMW but most nations produce their own maps at that scale (1 : 1 000 000) using variant symbols. There are some conventions about colour (e.g. blue for water, green for vegetation) which are widely followed but generally there is a great variation in the design and use of symbols. Therefore we can only look at general principles and outline the merits of the various choices open to map producers.

Topographical detail can be classified into points, lines, and areas (although in Chapter 7 below it will be found that computer data banks define everything in terms of points only). The distinction between points and areas is often confused by change of scale: an area on a large-scale map may reduce to a point at small scale. There is also a distinction between a line which is the limit of an area (e.g. the shore of a lake) and a line which represents a narrow feature (e.g. a footpath).

The variables for an area symbol are pattern and colour; for point and line symbols, size is also a variable; and for point symbols only, shape. Simple shapes and patterns are preferred since they require less drafting time.

Colour theory is discussed below in Chapter 5. The ability to distinguish differences in colour decreases as the area coloured decreases; hence colour is less useful as a variable for point symbols than it is for area symbols. In monochrome maps, of course, colour variation cannot be employed at all.

Fortunately they are usually at large scales where symbolization is least necessary. Occasionally monochrome small-scale maps are required to illustrate books or reports and these present special problems in symbol design and use.

Map makers in bygone centuries used pictorial symbols: small drawings of mountains, buildings, trees, etc. This approach is easily understood by the unskilled map reader and is still used in designing some symbols. Whales and mermaids are no longer added to nautical charts, nor are elephants and lions scattered across the less developed parts of Africa, but such features as lighthouses or wind-pumps can well be indicated by mini-pictures of these structures.

2.2.1 Point symbols

See fig. 2.1. The variables are:

(*a*) size, and here it should be noted that we can vary both the overall dimensions and also the thickness of line in 'open' symbols.

(*b*) shape, e.g. square, triangle, circle, cross.

(*c*) pattern, e.g. a solid square, an open square, an open square with diagonals, etc.

(*d*) colour (see above).

Fig. 2.1 Point symbols (enlarged)

A common and useful adjunct to point symbols is a set of conventional abbreviations, e.g. PS = police station, PO = post office, Ch. = church, etc. Use of these avoids the necessity of designing special symbols for such buildings.

2.2.2 Line symbols

The variables are:

(*a*) size: the width of a line or lines and the spacing between double lines.

(*b*) pattern: single or double or multiple lines, continuous or broken lines, lines of dots, dashes, crosses, circles, etc. crossbars or bars on one side only (see fig. 2.2).

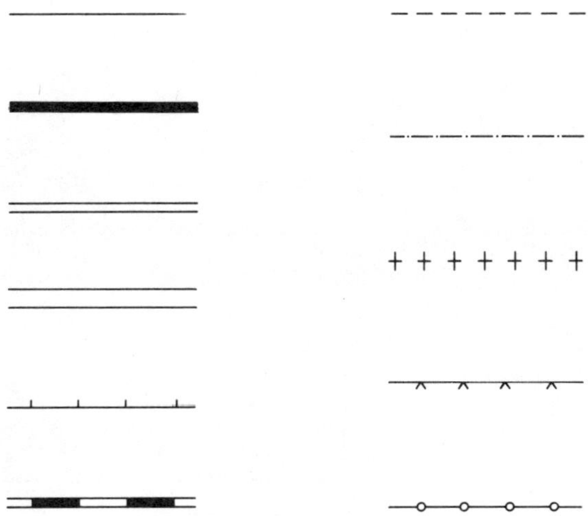

Fig. 2.2 Line symbols

(*c*) colour: common conventions are black for railways, red for roads, blue for waterways, brown for lines of equal elevation (contours).

Line patterns generally require extra time and labour in drafting and their use is increasingly avoided.

2.2.3 Area symbols

The simplest and most effective method of differentiating areas is to use different colours. Pattern may be either pictorial, e.g. a forest area is shown by an array of mini-drawings of trees, or geometrical, i.e. an array of dots or rulings. Dots may be varied in size and spacing, while lines may be varied in width, spacing, and direction (see fig. 2.3). However, in practical terms, a coarse geometrical pattern may clash with other map detail in the area and should not be used, while a fine pattern has the same appearance as a tint of the colour in which the pattern is printed.

Problems may arise if two systems of conventional colouring are applicable to the same area, e.g. a green vegetation symbol and a brown altitude symbol. In such a situation it is better to change one system from an overall area symbol to a boundary symbol, e.g. instead of showing a forest by an overall green tint, it may be defined by drawing a green line along the forest boundary. This method is more effective if the boundary line is emphasized by adding a strip of colour tint. In the above example a strip of pale green colour may be added on the forest side of the line. As a further improvement the strip

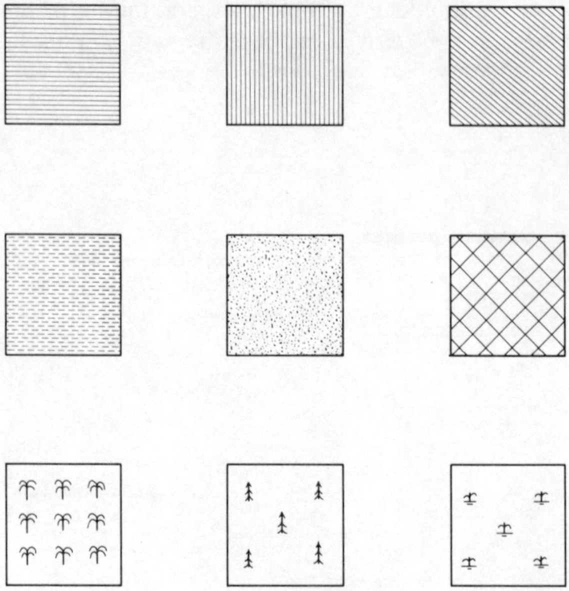

Fig. 2.3 Area symbols

may be *vignetted* (which means that the strength of colour across the strip will fade from dark green along the boundary line to pale green in the middle of the strip, and finally to no green) (see fig. 2.4). The process of adding symbols, tints, vignettes, etc., to the drawings is explained in Chapter 5.

2.2.4 Preprinted symbols

Where extensive areas have to be filled with a symbol (other than a colour tint) the quickest and most satisfactory method is to use preprinted symbols, prepared to a uniform size and spacing. The symbols are printed on sheets of transparent foil and may either be cut out and fixed on to the relevant areas of the drawing or be used to transfer an image to the drawing by a photographic process (see also Sections 2.6.5 and 5.7).

2.2.5 Symbols on monochrome maps

If a map has to be produced in only one colour there are special problems with line and area symbols. Area symbols should be avoided as they obscure other detail. If areas have to be identified this should be done showing only their boundaries. The least intrusive overall symbol, if one must be used, is a light grey tint made by a pattern of fine dots.

Different types of line detail have to be shown by varying line patterns, which increase drawing time. The greatest length of line symbol on a topographical map is usually the contour lines indicating heights. On a

Fig. 2.4 Vignette

monochrome map, contours are best omitted and some other method used to show heights.

The representation on the map of various features may now be considered in more detail.

2.2.6 Built-up areas

See fig. 2.5. Isolated buildings or small built-on areas are best shown solid black. Larger areas may be bordered in black or grey and filled with a grey or yellow tint. Within such a tinted area important buildings may be picked out in solid black.

On small-scale maps, villages and small towns are usually shown conventionally, the preferred symbols being circles of various sizes and patterns. For large towns it may be possible to draw the true outline of the built-up area.

2.2.7 Railways

See fig. 2.6. Since a plain black line is the easiest detail to draw, this should

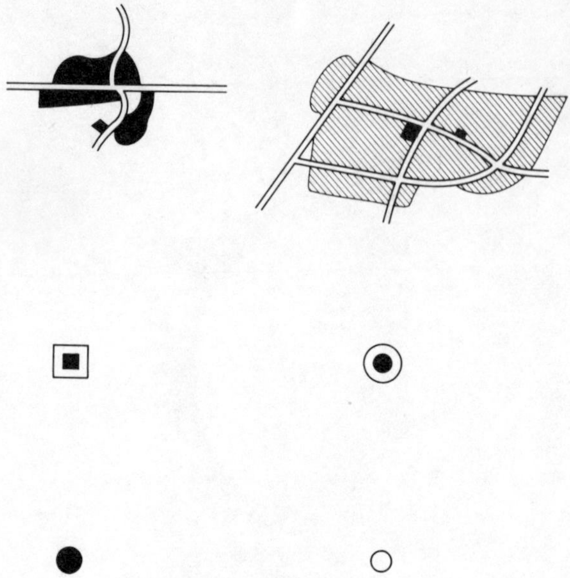

Fig. 2.5 Built-up areas

be the symbol used for the most widespread class of railroad in a country. This might be metre gauge single track as in East Africa, or double track 1.435 m gauge as in Western Europe. A more complex symbol can be used for the shorter lengths of other classes. For treatment of tunnels, see Section 2.2.10.

2.2.8 Roads and tracks

For economy in drawing it is best to avoid pecked lines and diced fillings. Trunk roads may be shown by heavier line or greater width between double lines; bound surface might be indicated by red filling, loose surface by brown filling, seasonal roads by no filling, tracks by a single red line.

2.2.9 Airfields

At small scales a point symbol is used, e.g. a pictogram of an aircraft. At medium scales the runways can be shown at correct scale and orientation. The runway surfaces (bound, loose, grass) can be indicated by colour (e.g. red, brown, green).

2.2.10 Pipelines and tunnels

Oil, gas, and water pipelines, some of great length, are now widespread and may be above or below the ground surface in various sectors. Some topographers argue that only visible features should be shown on a map, but the location of underground features is often important to some map users, and there is a

Fig. 2.6 Railways

good case for showing the alignment of buried pipes and tunnels. The usual convention is to show them by pecked lines.

2.2.11 Survey control points

In drawing a machine plot, photogrammetrists only show ground control points which have been identified and marked on the air photos and used to control the plotting. There are frequently other surveyed points which are pillared or otherwise permanently marked on the ground and useful to surveyors and others. A well-spaced selection may usefully be included in a map.

2.2.12 Water features

The accurate representation of rivers presents several drafting problems. The first is that rivers tend to change course, particularly in flat areas, each flood season. The width should as far as possible be shown to scale but for single-line rivers this can be so time consuming that it is better to adopt a system of uniform line widths, increasing downstream. Change of width can be made at suitable points such as where two streams join.

When the river is wide enough to be shown by a double line the space must be filled with a blue tint (for seasonal rivers a brown filling might be used). Generally in developing countries no complete survey has been made to locate the limits of seasonal and perennial flow, so no differentiation can be shown. A seasonal single-line river could be shown by a blue peck. Another symbol is required for watercourses whose position is uncertain.

Small lakes may be shown solid blue, but the usual convention for lakes is a solid blue line for the shore and a blue tint filling. Since there is no detail within a lake which might be obscured by the filling, there is no good reason to replace it by a vignetted edge.

The water level in some lakes may vary greatly and it is useful to distinguish permanent water from occasional coverage by using two blue tints.

It is desirable to enter the height of the lake surface above sea level, and if this is not constant, to add the date of the photographs from which the shoreline was plotted.

Areas liable to flood or seasonal swamps require a symbol: possibilities are horizontal parallel pecked lines, or a mixed blue and brown tint. The limits of such areas are usually ill-defined so the supposed edge need not be outlined.

Permanent swamps can be shown by adding a green vegetation symbol over a blue water tint. Some current maps show a blue vegetation symbol on a green water tint, which is illogical.

2.2.13 Vegetation

This is shown by a green tint or green symbols. Green symbols have less visual impact than black symbols; hence if it is desirable to distinguish readily between different types of vegetation (e.g. palms, bamboo, dense forest, low bush, etc.) black symbols on a green tint background may be used.

Plantations are best shown by a coarse green ruling (representing bushes or trees planted in lines) with black indicator letters (e.g. T for tea, S for sisal, etc.). This is simpler than designing a special symbol for each crop.

2.3 Boundaries

Boundaries are important in human affairs but their position on the ground is frequently far from obvious. They may be minutely described in words in documents but relating the description to actual position on the ground can be a frustrating and ambiguous procedure. A map has a special value as a medium for illustrating the position of boundaries.

Boundaries might be loosely classified as international, central government, local government, departmental or property. How many of these are shown on any one map depends on the scale and purpose of the map. On large-scale maps there is space to show all classes of boundaries whereas at small scales only some of the central government divisions (national, regional, provincial, county, district, etc.) are likely to appear.

Departmental boundaries such as police divisions, education, and health authority areas, etc. are unlikely to appear on general topographical maps; but forest reserves, national parks, nature reserves, etc. are usually shown.

Special boundary maps are sometimes produced which may be classed as thematic maps. An alternative is to make a drawing of the required class of boundaries (e.g. local government boundaries) and overprint it in a distinctive colour (e.g. purple) on to a standard topographical map.

Important boundaries should be printed in black and if they divide areas of equal status they should be narrow lines to allow maximum precision of position; to make them more prominent, a broad band of colour, usually red, is added symmetrically over the black line.

If the boundary is a limit, dividing land of unequal status, a different treatment is possible. For example a national park might be considered as higher status than adjoining land. Then a colour band, perhaps green, uniformly tinted or vignetted, may be drawn along the park side of the boundary.

Boundaries on the ground may be defined or undefined. Defined means there is some continuous visible physical feature which is related to the position of the boundary. It might be the Great Wall of China or the Berlin Wall or some minor work such as a road, hedge, ditch or bank or a natural feature such as a river bank, lake shore, or cliff.

Undefined boundaries are generally straight lines joining known points such as hill tops, river junctions or boundary pillars. The drawing of such boundaries on maps presents few problems.

The main problem in showing defined boundaries on maps is in deciding where to place the boundary symbol. If it is superimposed on the image of the defined feature, there is confusion, e.g. a black and red boundary symbol on a black and red road. To place the boundary on one side parallel to the feature is misleading, as is also the alternative of drawing short lengths on alternate sides. There are occasions when the simplest solution is to annotate the line, e.g. 'boundary is north bank of river' or other appropriate description.

2.4 Relief

2.4.1 The third dimension

The preceding paragraphs have discussed the transfer of information having two horizontal dimensions on the ground to the two horizontal dimensions of a flat map sheet. But ground detail has a third dimension: height or elevation (the term altitude is preferably used only in reference to height *above* the surface, as for aircraft). The word 'relief' is used to refer to the general state of non-flatness. One of the major problems in topographical map making is to arrive at the best way of showing ground relief on a flat map.

It is of course possible to produce three-dimensional maps by plastic modelling; but they are costly, awkward to store, and difficult to transfer information to or from. Also, unless the vertical scale is greatly exaggerated, they do not give such a good impression of relief as might be expected.

As mentioned above in Section 2.2, early cartographers used little drawings of hills or mountains to indicate upland areas. These were superseded by *hachures,* which were used, for example, on the early editions of the 1 : 63 360 map of Britain in the mid-nineteenth century. Hachures are lines drawn in the direction of greatest slope, thick at the top and thinning downwards.

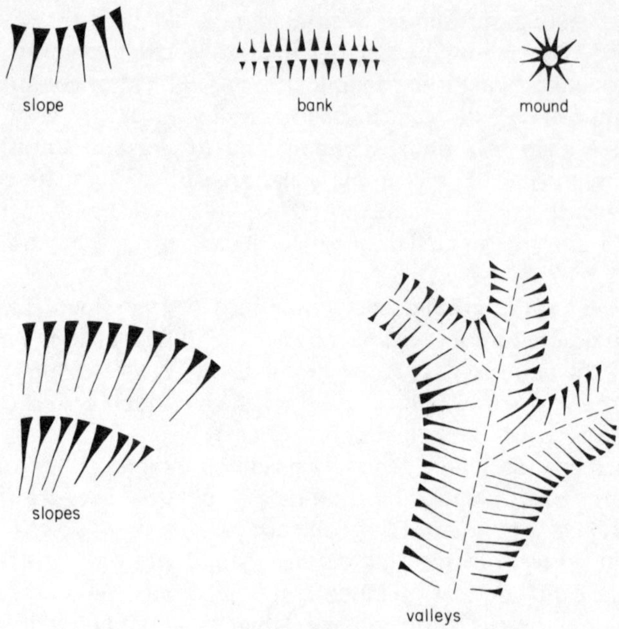

Fig. 2.7 Hachures

Closer spacing or thicker hachures indicate steeper slopes (see fig. 2.7).

On a small-scale map a mountain range shown by this method can look like a hairy caterpillar. Hachures inevitably obscure other map detail. However, they may still be a useful way of indicating relief on a monochrome map and they are used on current medium-scale maps to show small relief features such as mounds or embankments.

The next development was to abandon lines down the slope in favour of horizontal broken lines across the slope. These can be spaced more widely and obscure less of the other detail. Closer spacing indicates steeper slope and the more lines, the higher the hill. In the days when all mapping was done on the ground, skilled topographers sketched in *form-lines* directly on to the map field sheet; now they can be sketched by a photogrammetrist from the photostereo image on to the machine plot. They have no numerical values and no pretensions to being truly horizontal or at equal vertical intervals.

2.4.2 Height data

The systematic and accurate portrayal of relief could not develop until an adequate network of heighted points had been fixed on the ground, a branch of land surveying that tended to lag behind planimetric surveying. Vertical angles observed by theodolite have to be corrected for curvature and refraction, aneroid readings must be adjusted for temperature and humidity and are not very accurate, lines of precise spirit levelling tend to be far apart.

However, given a suitable number of heighted points identified on air photos, a photogrammetrist can rapidly extend adequate height information over a whole mapping area.

The datum level for heights on land is usually local mean sea level. The unit for heights and depths is now normally the metre, with the important exception that for air charts it is the foot.

2.4.3 Spot heights

For mapping purposes all scattered heights fixed by instrument may be grouped under this heading. In descending order of accuracy they are levelling bench marks, triangulation and traverse stations, aneroid heights, and photogrammetric heights. These are usually entered on the map to the nearest whole metre although this may not always be quite accurate.

Spot heights chosen to be shown on maps should be in positions where they are most useful: hill tops, lake shores, and at intervals along ridges and valleys.

They give accurate information, but no complete picture of the shape of the ground.

2.4.4 Contours

Contours are lines of constant elevation. On the ground they are of course invisible and imaginary but they are real lines in the map. On medium- and large-scale maps they are normally spaced at a constant *vertical interval* though this interval may be increased in mountainous areas with steep slopes.

The most suitable vertical interval must obviously vary with scale in order to obtain a convenient average horizontal spacing. Crowded contours are difficult to read and tend to obscure other detail whereas wide spacing reduces the amount of height information which could usefully be shown.

For contours based on foot units there was a good rule-of-thumb for the most convenient vertical interval in terrain of average relief:

scale:	1 : 50 000	1 : 100 000	1 : 250 000	
vertical interval:	50 ft	100 ft	250 ft	etc.

i.e. the vertical interval is the number of thousands in the scale fraction. In flat country the vertical interval might be halved and in mountains doubled.

This rule does not conveniently convert to metres. In the above examples 10, 20, 50 m might be suitable in flat lands or 20, 50, 100 m in hilly lands but neither of these may be the most suitable for average terrain.

In choosing a set of contour values it must be remembered that when derived maps are made at smaller scales from basic maps it is essential to select and use existing contours and avoid having to plot or interpolate additional ones. In the above examples 100 m contours for the 1 : 250 000 map can be compiled from both 50 m contours on the 1 : 100 000 map and from 20 m contours on the 1 : 50 000 map but 50 m contours could not be compiled from 20 m contours because alternate 50 m are missing in the latter series.

To assist identification of contours on the map, every fifth contour may be drawn thicker and is called an *index contour*. For example, if the vertical interval is 20 m, every multiple of 100 m will be an index contour.

Contour accuracy

If contours were precisely plotted the height of any point on the map would be determined with an error not exceeding half the contour interval. For example, a point lying between the 20 m and 30 m contours could be assigned a height of 25 m with an error not exceeding 5 m. However, contours are not so precisely plotted; the photogrammetrist aims to keep the error in a contour to less than half the contour vertical interval. Hence in an extreme case the error in an interpolated point could be as great as the vertical interval. If the photogrammetrist cannot limit error to half the vertical interval, either the latter must be increased or the contours must be downgraded in classification to form lines.

Contour numbers

Contour values (heights above sea level) must be added at convenient positions all over the map. A south-facing slope should be selected and the values added like a ladder up the line of greatest slope. The top of a number should be on the uphill side and should be readable (like all other map alphanumerics) from the south. It is standard practice to break the contour line and add the figures centrally in the gap.

Depression contours

Closed-ring contours normally indicate that the ground within the ring is higher, but it could in fact be lower, e.g. a hill of volcanic origin may have a crater in the top. The direction of slope is known from the contour numbers, but the innermost ring can be too small either to accept a number or to leave space for a central spot height. If there is an inward slope the convention is to add ticks on the lower inner side (fig. 2.8).

Contours give a reasonably accurate and fairly complete indication of relief. They can be used to assess slopes and gradients and limits of visibility. They do not give a quick overall impression of relative height; a glance at a contoured map does not immediately show where is the highest or lowest area. To provide such a picture it is necessary to resort to a further convention.

2.4.5 Layer tinting

A layer is a uniform colour added to all land between two selected contours,

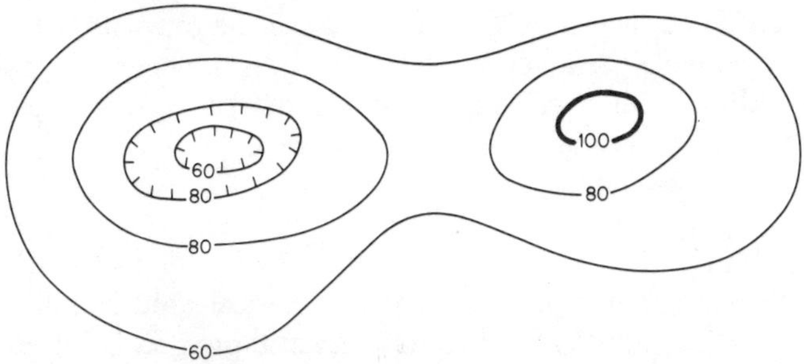

Fig. 2.8 Contours, index contour and depression contours

e.g. all land between 100 m and 200 m above sea level might be coloured pale buff and between 200 m and 300 m dark buff, and so on. Relative heights over large areas are then immediately apparent, although relative heights within one colour zone are not. The method is not used on large- and medium-scale maps, because one sheet, covering a relatively small area, might lie entirely within one colour zone.

Vertical intervals

There are much larger areas of the world at low elevations than at high ones and a small change in elevation in a low area is more significant than a large change in a high area. For this reason, the vertical interval covered by one layer should be increased as height increases. For a ten-layer map the limits of successive layers might be at 0, 50, 100, 200, 400, 600, 1000, 1500, 2000, 3000, 5000 m.

Choice of colours

Each layer must be distinct from each adjoining one but without too sharp a contrast. The change of colour in a series must be progressive. The colours must be light, not to obscure map detail. They must not clash with other coloured area symbols, e.g. green vegetation symbols and they should not conflict with common mental associations, e.g. green should not be used over desert areas. For economy in printing it should be possible to produce the chosen series from tints or combinations of a minimum number of inks.

Tints should not be printed over water areas where the layer colour might modify the blue. Water areas are level and therefore fall entirely within one layer zone.

In the early days of layer tinting, the colour succession was based roughly on actual land surface colour in a temperate climate, ranging from lush green at vegetated low levels through the browns of higher barer ground to purple for heather and white for snow on the mountains. A more satisfactory succession likely to be found on recent maps would pass from yellow through various shades of buff and brown perhaps finishing at the top with greys (but some countries use grey for land below sea level). On a monochrome map only grey tints could be used; but layer tints are rarely required on such maps.

For the preparation of layer tints, see Chapter 5.

2.4.6 Hill shading

Whereas contours and layering have numerical values, hill shading has evolved from hachures and has no quantitative basis. There are two approaches to the technique. For the first we imagine a relief model of the terrain illuminated from the northwest. Then south and east facing slopes will be shaded, shadow increasing with steepness. This gives little information about the steepness of north and west facing slopes. The second approach imagines vertical illumination, under which reflection decreases (and hence shade increases) as slope steepens, irrespective of aspect.

Shading is applied to the drawing by crayon or spray and a good result requires much skill and experience by the operator. Shading inevitably (like hachures) must to some extent obscure other detail. It is debatable whether it is cost-effective.

2.5 Geographical Names

Each time an area is mapped at a larger scale than any new previous mapping, i.e. there is a new basic mapping, the new map will contain a lot of detail not previously shown on any map, and much of this new detail will be known to the inhabitants of the area by local geographical names which should also appear on the map.

The *collection* of such names is a field task for a surveyor or cartographer, usually at the field completion stage of producing the map.

Names must be recorded systematically, preferably on a form designed for the purpose. Places in different major administrative units should be listed on separate forms.

The heading on a form should indicate location and map number and the information given for each name should include grid reference, nature of feature, local spelling, language, meaning, etc.

The collector must be sure that each name is linked to the correct feature, e.g. in Kenya there is an area called Suswa, but on the map the name is wrongly applied to a mountain in the area. Also he must check that each name is a *specific* name for a particular feature and not just a *generic* name for the class of feature, e.g. in Galla, galana means river. But there is a River Galana on the map.

The completed forms should be checked and amplified by a local authority familiar with both the geography and language of the area. Spellings recommended for use on maps should be confirmed by a national names authority if one exists.

Frequently there is no such authority and the map production agency has to make its own decisions. It is therefore pertinent to state here some general principles.

2.5.1 Alphabet and Orthography (system of spelling and pronunciation)

These are generally those used for the official language of the map-producing country. If there is more than one official language, one must be selected for map use or alternative spellings must be given (e.g. in Belgium the Flemish and French spellings for the same place are Gand and Ghent and both are official).

2.5.2 Foreign names

In all continental states, national mapping cannot avoid covering a small or large extent beyond the national frontiers, including areas where names may be spelt in a different orthography. The rules often have to be flexible. For instance Kenya maps may include Chisimaio in Somalia, a relict Italian spelling. Should it be printed in that form or should it be transliterated into Swahili (which is the official language for spellings on Kenya maps)? It then becomes Kisimayu which is immediately recognizable by Swahili readers as meaning ‘the high island’. Generally, established foreign names, whether within or without the national boundaries, are better left as they are (e.g. Beaulieu and many other place names of French origin in England, where they are now pronounced quite differently from the original French) unless they can be replaced by an authentic local name (as Muranga replaced Fort Hall in Kenya).

Once a name has appeared on a map it becomes increasingly difficult to change since it may be repeated in timetables, telephone directories, gazetteers, etc. Hence map producers have a special responsibility to show the most acceptable spelling.

2.5.3 Vernaculars, generics and specifics

There may be within a country *vernaculars* which have a different orthography from the national or official language. Kenya has several, but this is not a problem confined to developing countries — in Britain in recent years many spellings have been changed in Wales from English to Welsh forms.

It is probably better to accept as far as possible on maps the vernacular spelling although a non-native of the region will frequently mispronounce it. This is true in Wales (e.g. Llanfair or Foel Uchaf) and also true in Kenya. For example, Thika is so spelt in Kikuyu orthography, but the Kikuyu pronunciation rendered in English or Swahili script is more like Dheka. If the latter spelling is used on the map, then the Kikuyu people who are possibly the majority users of the map, will not recognize it as the name they write Thika. However if the vernacular alphabet includes letters not found in the official map language, a transliteration is unavoidable, e.g. there is no Q in Swahili so Qadaduma (a place on the Ethiopian border) has to be rendered Gadaduma.

For names consisting of more than one word it is desirable to avoid mixing languages. In most such names one word is a *generic* indicating the type of feature (e.g. river, hill) and the remainder is *specific*, being the particular name of the individual feature. So Engare Nairobi is preferable to Nairobi River, and Got Abuor to Abuor Hill. Lack of understanding of the generic parts of names and their meanings often leads to unnecessary repetitions on maps, e.g. Mount Kilimanjaro, Sahara Desert, Lake Nyasa, The Algarve (Kilima = hill, sahara = deserts, nyasa = lake, al-garve = the west).

2.5.4 Glossaries

Therefore on maps containing many foreign or vernacular names it is helpful to add a *glossary* (literally a word-list) of all the generic terms attached to names not originating from the production language of the map, e.g. the following are only a few of the words meaning 'hill' which might be found on a map of eastern Africa: bur (Boran), got (Luo), jebel (Arabic), endoinyo (Maasai), murua (Turkana).

On topographic maps symbols are used for many generics, e.g. a blue line indicates a river. This may allow the cartographer to dodge the problem of finding the correct generic form by omitting it. It is sufficient to add the specific, e.g. Nile and omit the generic River. But if the specific is an adjective then logically it should not stand alone without a generic, e.g. Engare Nairobi = cold river, and it is insufficient for the map to show only Nairobi (= cold).

Generics are often shown by an abbreviation, e.g. Red R. for Red River. The abbreviation is a class of symbol and should be shown in the reference panel.

It is a convention in mapping to omit from names all kinds of accents, punctuation, and diacritic marks. This is because a hyphen or stop or other mark can be mistaken for a building or other map detail.

2.5.5 Gazetteers

Any atlas, containing many pages of maps, normally includes an index, called a *gazetteer*, of all geographical names printed on the maps, with a latitude and longitude or grid reference.

A gazetteer originally meant the editor of a newssheet which sold for the prize of a gazeta, an old Italian coin. An editor dealing with foreign news needed a reference book to check up on foreign places and such a book in turn acquired the name of a gazetteer.

A single-sheet map does not warrant the addition of a gazetteer, but a full series of basic mapping, probably at 1 : 50 000, certainly does. The total number of names appearing on a map will vary with the area of land covered, the density of human settlement, the scale of the map, and the amount of effort put into collecting names during field completion. On a 1 : 50 000 map a range between one name per kilometre square and one per 10 km square may be expected.

A gazetteer is published in book form with the names listed in alphabetical order (this can produce problems; e.g. Odera's Bridge in Luo is Ulalo Kodera, the K being a possessive prefix. Should it be indexed under U, K, or O?). Each name should be followed by a code or abbreviation to indicate its topographic feature, e.g. mt for mountain, br for bridge, its position coordinates, and another code for the political or administrative unit in which it is situated. The volume must contain a key to code letters, and preferably a topographic glossary and an index map showing the boundaries of the administrative units and location of the vernaculars.

A gazetteer is only an abbreviated record of geographical names. A full record should also be kept by preparing a card for each name. On each card should be entered all the information from the field collection sheet and also references to any correspondence or decisions about the spelling of the same name on different maps. The cards are kept in alphabetical order in card-index drawers. From each card a punched computer card can be made. The computer is programmed to sort the cards into alphabetical order and make a print-out of the required information which can then be converted into book form.

2.6 Letterpress

The alphanumerics on a map are collectively referred to as type or letterpress. Freehand lettering and numbering is now rare as also is the use of stencils. Preparation of type is largely mechanized.

2.6.1 Type variables

See fig. 2.9. A single printed letter can vary in size, shape, and various other characteristics which must each be specified if the cartographer is to obtain exactly what he wants to appear on the map.

Case: capitals and small letters are known as *upper case* and *lower case* respectively.

Italic: normal lettering is upright and this is often called *Roman*. This can be confusing because the term Roman is also used to distinguish our alphabet from Cyrillic, Greek, Arabic, and other alphabets. Sloping lettering is called *Italic*.

Serifs: these are the small lines attached to the extremities of letters which are said to improve legibility, though they can be regarded as only ornamental. Alphabets without serifs are called *sans-serif*.

Weight: the terms *bold*, *medium*, *light* are used to indicate thickness of line; normal type being medium.

UPPER	lower
Roman	*Italic*
Serif	Sans
bold	Light
Open	SHADED
condensed	**extended**

Fig. 2.9 Type variables

Set: the width of an average letter (such as a, c, e, n) is usually 70–80% of its height. Narrower and wider letters are respectively called *condensed* and *extended*.

Open/Shaded: large letters normally have proportionally thick lines, which can obscure a lot of map detail. To avoid this, the lines of *open* type are clear in the middle while *shaded* type has only a light grey filling.

Fount: there are hundreds of type designs in daily use, and *fount* is the general term, covering both designer and maker, used to identify them. Some founts in common use on maps are: Gill, Times New Roman, Univers, Futura.

Point: size of type is usually defined by the height of a capital letter, which could be stated to a precision of 0.25 mm; such metric measurement is already in use and likely to be standard in future. However, measurement by points is still widely employed by printers and it is necessary for a cartographer to comprehend it.

A size quoted in points is not a measure of the height of the actual letter but of the metal block backing the letter, as used in traditional letterpress printing. The letter may occupy any proportion of the 'shoulder' (face) of the 'sort' (metal block). One inch is roughly 72 points but a 36 point capital is not half an inch high because the 36 points includes the 'beard' which is the space between the base of the letter and the top of the next line of type. To be sure of getting type of a desired size it is necessary to choose it from a manufacturer's catalogue.

The limits between small, medium, and large type are generally set at 8 and 24 point. Sizes in use on maps range from 5 point (capitals 1.7 mm) for the least important detail up to perhaps 48 point (16 mm) for the sheet title. In practice, for quick and easy reading, 8 point is the smallest useful size on maps, and the use of smaller type should be avoided as far as possible.

2.6.2 *Choice of type*

The map specification (see Section 2.8) must state exactly the details of the type to be used for each class of map information. For example, it might be decided to use Times New Roman upright capitals for main towns (the point size varying with size of town), Univers medium upper and lower case 8 point for villages, Gill Sans medium condensed upper and lower case 6 point for railway stations, etc.

Names in bold type or wholly in capitals take up extra space without being easier to read than medium or upper and lower case; use should therefore be restricted to a minimum.

2.6.3 *Ordering type*

This is a task which must be done systematically. Initially all the names are probably on one monochrome compilation sheet. The line-work for each colour is then drawn on a separate sheet or flap. Consider the blue flap. A combined print of this drawing and the grid should be made, and on this entered all the names and figures which are to be printed blue. They can then be listed with their grid-square numbers (which will aid placing the printed names when they are received). On the type order form names should be grouped by fount, point, case, etc.; this saves time in the letterpress section.

2.6.4 *Positioning type*

Names should be placed close to detail to which they refer but should not obscure other detail, either on the same flap or on any other flap. A name is normally placed parallel to the top margin (i.e. east-west on most maps) readable from the south, but for irregular line detail like rivers the name must run along the line of the river; similarly for names of features which have a long axis running in some direction other than east-west. Sometimes a name has to be cut and the letters arranged around a curve.

Names are easier to find and read when the initial capital letter has a clear space round it; this implies that they should not be placed immediately to the right of features. But this position is the one which indicates most strongly the link between the detail and its name. Map-user requirements may influence a decision on which attribute is the more important.

2.6.5 *Preparation of letterpress*

Movable metal type is still used for preparation of small quantities of type, being set by hand in small handstamps for stamping directly on drawings or in small machines for printing on plastic foil (which is then stuck on the map). But for quickest preparation and best results a phototypesetting machine is used. There are several makes available.

A common design has a set of circular discs, each disc being for a particular fount. On one disc are all the upper and lower case letters and figures and some common symbols, all in negative form (transparent letter in opaque surround) and of the same point size (e.g. 18 pt). The optics of the machine can be adjusted to enlarge or reduce the letters to any required common point size.

A letter is selected on a keyboard or index panel, and on pressing a switch, light passes through the negative letter to make an image on a sheet (usually A4 size) of transparent sensitized plastic foil. On development, the names appear positive on the foil. The foil sheet can be automatically moved to give required spacing between letters, words, and lines. The foil is very thin and supported on a thicker backing material.

If the machine has a step-and-repeat facility, it may be used for preparing sheets of preprinted area symbols (see Section 2.2.4).

The use of small computers or 'word-processors' for type setting is also rapidly developing.

2.6.6 Mounting letterpress

The printed names are cut out of the sheet and the foil detached from the backing. If the foil is not self-adhesive then wax or other adhesive must be used to attach it to the fair drawing. If the drawing is reverse reading (as is a positive made by contact from a scribed sheet) then the name must be mounted in reverse.

2.7 Marginal Information

The preceding paragraphs have been mainly concerned with the actual map, the graphic image of a part of the Earth's surface. On the finished published sheet this area is usually enclosed in a fine sheet line or neat line. To make the map sheet a complete and useful document it is necessary to add a variety of other information.

On a single-sheet map, the mapped area often has irregular boundaries and does not reach the neat line; space can be found within the neat line to add the required reference information.

2.7.1 Standard margin

If the mapped area is covered by a series of sheets then there will be no space within the neat line on most sheets. For such a series it is usual to design a *standard margin* which can be used with a minimum of alteration for all the sheets (see also Section 1.11.1).

Outside the neat line is an outer margin or border, usually a thick line. In the space between the two lines can be placed the following information which is related to the map detail:

Grid values, graticule values, off-sheet destinations of railways and main roads, parts of names of large areas which extend beyond the sheet edge.

The next group of items are placed above the border in the top margin.

2.7.2 Sheet name

This is placed centrally at the top in large type. For rapid location of the mapped area every sheet should have a name, which is usually that of the largest town or other most prominent feature in the area.

2.7.3 Series number

A map may be a single sheet or a series of sheets. However this may be, every map should have a publisher's serial number for rapid identification. A map sometimes has more than one series number if it is published at different times (by mutual agreement) by different map production agencies.

2.7.4 Series name

A series name, like a sheet name, is for rapid identification of the area covered by the series. If the same area is covered by another series at a different scale, then the scale may be included with the series name, e.g. East Africa 1 : 250 000.

2.7.5 Sheet number

A map series with many sheets should have a logical numbering system. Thus an IMW sheet number starts with N or S to indicate the hemisphere, then a latitude index letter (A = 0–4°, B = 4–8°, etc.) followed by a longitude index figure (1 = 180–174W, 2 = 174–168W, etc.).

If the sheet lines of series at larger scales fit into the IMW pattern, sheets may be numbered by adding further letters and/or figures to the IMW number. Thus the sixteen 1° × 1½° sheets of a 1 : 250 000 series falling in IMW sheet SA36 might be numbered SA36A, SA36B, etc. or SA36-1, SA36-2, etc.

However, this system becomes cumbersome at larger scales and it is more convenient to start again on a national basis. For example a country may be divided into ½° squares, with serial numbers starting from 1 in the northwest corner of the country. Then a 1 : 100 000 sheet filling one square might be sheet 99, a 1 : 50 000 sheet ¼° square could be 99D, one of sixteen 1 : 25 000 sheets occupying the same ½° square could be 99-15, and so on.

2.7.6 Edition number

It is not sufficient to identify a map by its geographical location and scale; it must also be identified in time. When stocks of a topographical map sheet are exhausted they can be replenished (provided the series has not become obsolete and is not being discontinued) by reprinting from existing drawings, or more commonly, the drawings are first revised to reflect changes on the ground since the previous edition. In this case, the edition number must be advanced. If the map is subject to a joint production agreement between two production agencies (e.g. it might be a sheet lying across an international frontier) then either agency might produce the new edition and this should be indicated after the edition number.

2.7.7 'Refer to' box

To aid rapid and adequate identification many maps have a panel showing the series number, sheet number, and edition. This is often prefixed with 'Refer to this map as', hence the title of the panel or box. Consider this example:

Series Y503
Sheet SA37-5
Edition 3 SK

This would identify the map as a 1 : 250 000 scale topographical map covering the Nairobi area and published by Survey of Kenya.

The preceding items are normally found in the top margin. Other marginal information must be positioned and arranged according to the sheet layout design. Any of the following may be shown.

2.7.8 Scale

This may be stated as a ratio or fraction and also shown as a linear scale (see fig. 2.10) which may be in kilometres and metres only or may have additional bars to show miles or nautical miles.

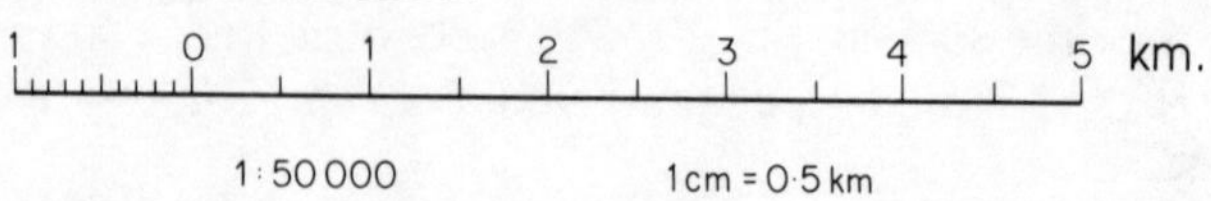

Fig. 2.10 Scale bar, scale fraction, scale statement

Note that if the size of a whole map is altered by photographic reduction or enlargement, the altered linear scale is still valid but a ratio or scale statement needs amendment; e.g. a 1 : 50 000 map reduced to half-size has a scale of 1 : 100 000.

2.7.9 Height information

This includes a statement of the unit (metres or feet) and of the contour vertical interval, an index to the layer tints, numbered in metres and feet, and if there are no layer tints, a vertical scale bar showing conversion between metres and feet.

2.7.10 Conventional signs or symbols

These are usually grouped together in a 'reference box' and should include examples and explanations of all symbols and conventions used in the map. Including abbreviations there may be as many as a hundred so this is often the largest component of marginal information. However, the same box (and much of the other marginal information) can be used for every sheet of a series.

2.7.11 Sheet index

On a series map there must be a diagram showing the numbers of the adjoining sheets. If there are not many sheets in the series the diagram may represent the whole series in which case it can be used unaltered in the margin of every sheet.

2.7.12 Projection, grid and magnetic data

Information about the following may be given in a tabular statement: spheroid, projection, datum, grid origin, false origin, scale factor. This can be followed by an example of how to obtain the grid references of a point of topographic detail. A diagram is the best means of showing the relationship (usually the values at the centre of the sheet) between the meridian, grid north and magnetic north, together with a statement of the date and rate of change of the magnetic declination.

2.7.13 Boundaries

Major administrative units may cover large areas of a map (sometimes the whole map area falls in one unit). Then naming the unit becomes a problem because the letters must be large and widely spaced, whereby they tend to obscure other map detail and are not easy to read. It is therefore better not to name extensive areas on the face of the map, but to show only their boundaries. The names and outline boundaries can be shown in a small marginal index diagram generally referred to as the 'Boundaries box'.

2.7.14 Boundaries disclaimer

As pointed out in Section 2.3 above, maps are an important medium for illustrating the position of boundaries but there are many complications when paper positions have to be interpreted on the ground. Map makers therefore take the precaution of adding on any map that shows boundaries the disclaimer 'This map is NOT an authority on boundaries'. Nevertheless, surveyors not infrequently find themselves called as expert witnesses in court cases involving boundaries.

2.7.15 Passage warning

Wherever roads, tracks, or paths are visible on the ground, a topographical map should show them. But some may be private property and some may be more or less physically impassable after rain, snow or for other cause. It should therefore be standard practice on such maps to add a warning 'Passage along roads and tracks shown on this map may not always be possible or permitted'.

2.7.16 Reliability and/or Construction diagram

This is another marginal box showing the sources, dates, and reliability of the air photos and/or other material used for compiling various parts of the map.

2.7.17 Sheet history

This amplifies the previous information by listing the sources of ground control, the dates of field completion or revision, the larger scale maps used, the agencies who carried out the plotting and drawing, etc.

2.7.18 Publication, printing, and copyright

These are short notes with dates showing the printer, the number of copies, the publisher, and the owner of the copyright. The international symbol © is used in place of a lengthy statement about copyright.

2.8 Map Specification

It should by now be apparent to the reader that there must be a written specification for a new map, drafted before the drawing of the first production sheet. The main items to be specified in detail are: scale, sheet lines, paper size, projection, grid and graticule, margin layout, conventional signs and colours, line gauges, list of type, printing ink colours.

A specimen map (which may initially be a modified copy of one produced by another organization) should be attached.

2.9 Advisory Committee

As with any other manufactured article, it is very difficult for a map producer to design a product that satisfies all users, or even a majority of them (see comments on ground/air graphics at section 3.3 below). Many major mapping organizations therefore take steps to set up an advisory committee comprising representatives of the principal users of their products. This may have the top status of a National Cartographic Committee or may be simply a more humble map users' committee; whatever its title, it can be of great assistance in leading to optimum design.

'The user never seems to be satisfied with the map as it is and there are many ways he may want to change it.' J. Winearls (1974) *Proc. Am. Con. Map Librarians*, p. 12.

3

Special Maps

3.1

About 150 states are members of the United Nations but only a small proportion of them have an organization exclusively engaged in chart production (marine charts, lake charts, and aeronautical charts) so that many are either dependent on foreign agencies or local production is carried out by the same organization as produces topographical maps. It should therefore be useful to examine briefly the special features of charts and differences from standard topographical maps.

3.1.1 Nautical charts

These are also variously known as *marine charts*, *hydrographic charts*, or *admiralty charts*. They differ somewhat from *bathymetric* charts which are virtually topographic maps of the ocean floor.

3.1.2 Projection

Mercator is the most used projection because straight lines on the map are lines of constant azimuth on the ocean. Small-scale charts of large areas (e.g. oceans) are often on gnomonic which can be used for planning long voyages, since the shortest distance between two points is a great circle and great circles are straight lines on this projection. Some charts stated to be on a gnomonic projection are in fact on some form of conic on which straight lines are close to representing great circles.

3.1.3 Scale

Coastal charts are generally in the 1 : 50 000 to 1 : 300 000 scale range, while port charts are at larger scales. Such charts are rarely in series and scale is determined by the area covered and the chosen paper size. The details inland from the coast which may be shown on such charts are usually compiled from local maps (where these exist and are up-to-date) and therefore there are obvious advantages if it is possible to adopt scales for charts which are the same as standard scales of local land maps.

Mercator and gnomonic charts covering large areas of ocean have appreciable scale variations and it is usual to quote the scale at one or more standard latitudes.

3.1.4 Sheet lines

Since the Mercator graticule is rectangular, graticule lines form convenient sheet lines. Successive charts, whether at the same or different scales, usually overlap, in order to ease the difficulty of plotting courses which cross the border of two sheets. If there is a large-scale port chart falling within the area of a small-scale coast chart, the sheet lines of the former are shown on the latter, and little detail is shown on the coast chart within the area of the port chart. The general principle is that the navigator should always be using the largest scale chart available in any area.

3.1.5 Graticule

Ocean navigation out of sight of land is based on latitude and longitude fixed by astronomical or satellite observations hence the graticule is the basis for reference and measurement on the chart and there is no grid. In some busy waters there are systems of position fixing from radio stations and these require a special 'lattice' overprint on charts.

3.1.6 Orientation

A special feature of charts is a large 'compass rose' printed in some convenient space, graduated in single degrees. Zero points to true north. This compass serves as a protractor; azimuths and bearings can be quickly transferred to or from it by parallel ruler. A second rose with zero on magnetic north may be printed inside the first one.

3.1.7 Magnetic data

The majority of seafarers rely on a magnetic compass for direction and therefore a statement and diagram of magnetic declination and variation (annual change) are essential.

3.1.8 Units

Again, because navigation is based on astronomy, the basic unit of distance is one minute of latitude, called a nautical mile. The average value of 1852 m is an international nautical mile.

Depths of water are in metres (0.1 m in shallow water) for all new work although there are many older charts showing fathoms (= 6 ft). The metre is also replacing the foot for all heights above sea level.

Horizontal scale bars show nautical miles, cables (1 cable = 0.1 nautical mile), metres, and feet; vertical bars shown feet, metres, and fathoms.

3.1.9 Horizontal datum

Marine charts are based on points whose latitude and longitude have been fixed by astronomical observations. The survey of coastal lands is based on the local national datum. This may result in a discrepancy in the position of the graticule shown on coastal topographical maps and charts. For example, the datum for mapping in East Africa is the Arc of the 30th Meridian. At the Kenya coast the discrepancy between map and chart graticules is over 200 m. This may eventually be resolved by adoption of a world geodetic datum based on satellite observations.

3.1.10 Vertical datum

The land surveyor and cartographer measure and record heights above the local surface of the geoid, which is taken as mean sea level. The mariner needs to know heights and depths above and below the water surface *at any time* and this is nowhere constant: on lakes there are seasonal variations and at sea there are tides. This results in the use on charts of two datums, neither of which is mean sea level, and neither of which has a constant difference from mean sea level.

Probably the most important information supplied by a chart is the minimum depth of water at any point; hence for depths the datum (called *chart datum*) is the lowest normal water level, depths below which are called *soundings*. This datum is also used for heights in the intertidal zone between low and high water; the figures are underlined to indicate that they are measured upwards from the datum. Alternatively the word 'dries' may be added before the figure. These are called *drying heights*.

For chart detail on land and above the highest normal water level, that is the datum level. Twice a month the tidal range reaches a maximum and *spring tides* occur; high tides occur twice a day and in some places they are of unequal height; hence the datum may be variously referred to as High Water Ordinary Spring Tides, Mean High Water Springs, Highest Astronomical Tide or Mean Higher High Water. The height of this datum above mean sea level varies along a coast and particularly inside narrow inlets; hence there is no constant for converting land heights on charts to heights on maps and vice versa.

3.1.11 Paper

Navigators plot position fixes and courses in pencil on charts; such work is later erased. Charts should therefore be printed on tough paper. Plotting is usually done on a large table in a chartroom and sheet size can be larger than for topographical maps. Charts may be as large as A0. (84 × 120 cm) or even larger. Such large sheets are usually stored folded in half.

3.1.12 Colours

Charts, like maps, were formerly printed in monochrome — black ink on white paper. The use of colour on charts started long after it had become standard practice on maps. Modern charts use magenta for lights, radio beacons, cables, sea routes, and boundaries of sea areas (e.g. anchorages). Land is printed yellow. Shallow water (e.g. down to 5 m) is printed solid blue while the next significant isobath (depth contour) has a blue ribbon edging. Deep water remains white for navigational plotting.

A fifth colour can be economically printed by combining two of the above; printing blue on yellow gives a green which can be used for the intertidal area between the yellow land and blue shallows.

3.1.13 Detail

The detail shown on a chart is carefully selected so that everything useful to the mariner is shown and no more. The topography of the coastline is fully shown but inland only prominent objects visible from the sea, e.g. hills, radio masts, high buildings, chimneys, etc.

On or below water level are shown: (*a*) aids to navigation: lights, buoys, and beacons; (*b*) dangers: rocks, wrecks, and cables; (*c*) currents and tidal streams: direction and velocity; (*d*) nature of the bottom: sand, mud, rock, etc.; (*e*) routes and limits; (*f*) soundings and approximate isobaths.

Information useful to mariners which is not suitable for printing on charts is compiled in book form, such a book being called a 'Pilot' (or Sailing Directions). It gives information about ports and harbours and navigation along coasts, climate, and many other relevant details.

3.2 Lake Charts

The charting of lakes has been somewhat neglected; obviously any lake on which there is navigation, even if only by relatively small craft, should be charted. Generally, charting is a useful step towards developing fisheries, irrigation and hydro-electric schemes, flood control, etc.

Lake charts naturally use the same horizontal datum, projection, grid, and units as maps of the surrounding land.

3.2.1 Vertical datum

As with sea charts, the lowest water level must be adopted as datum. Although lakes have no tides, water level varies seasonally, particularly if a lake has no outlet. The elevation of the datum above mean sea level should be quoted on a lake chart.

3.2.2 Soundings and isobaths

Soundings can be spaced more closely on a lake than at sea so that submarine contours can be more complete and reliable. As at sea, they are measured downward from chart datum, and also as at sea, there is often need for a special tint over the area between datum level and highest water level with a special notation to show 'drying heights' above chart datum.

Soundings are more important in shallow water than in the deeps so that contour vertical interval should be closer near the shore. A typical series of values might run: 1, 5, 10, 20 m below datum.

3.3 Aeronautical Charts

An airman's view of the ground is much the same as that of an air camera which takes the photos from which topographical maps are plotted, hence one might suppose that these maps are perfectly suitable for use by aviators. Indeed in many parts of the world local topographical maps are the only medium-scale material available and are used for local flying. But there is always a demand for specialized air charts because, like the mariner at sea, the airman requires specialized information. A high percentage of the information shown on a standard topographical map is of little interest to him. Attempts have been made to produce a map/chart suitable for use both on land and in the air, one example of which is the Anglo-American Joint Operations Graphic at scale 1 : 250 000. Even the name *Graphic* is a compromise, and two separate editions are needed for ground and air use, neither of which is as satisfactory as a fully specialized map or air chart.

3.3.1 Scales

For in-flight navigation 1 : 500 000 and 1 : 1 000 000 are the most favoured scales and there is a world series at the latter scale. 1 : 250 000 is suitable for small aircraft on local flights. Approach and landing charts for individual airports are at larger scales. The latter are usually printed in monochrome (magenta) on small (A4) sheets.

3.3.2 Projection

Airmen are even more concerned than seamen to follow great circle courses. On the gnomonic projection these are straight lines but it is in many ways an unsuitable projection, particularly for series maps. The World Air Chart 1 : 1 000 000 series therefore uses the Lambert conformal conic with two standard parallels. On this projection long straight lines approximate closely to great circles. For larger scales like 1 : 250 000, the Universal Transverse Mercator or a local Transverse Mercator is preferred. Datum and sheet lines can be the same as for local topographic map series.

Navigation in the air, as at sea, is basically astronomical. Hence a graticule is more use than a grid, and a scale of nautical miles is included in the chart margin.

3.3.3 Vertical datum

International flying is governed by the International Civil Aviation Organization and seems to be firmly committed to the foot for altitudes, with little prospect of an early change to the metre. This requires that all spot heights, contours, layer tints, etc., must be in feet and is a major obstacle to joint ground/air maps.

3.3.4 Chart detail

Specifications for air charts should follow international standards. Local air information can be obtained from the regional or national civil aviation authority which should also check the proof before the chart is finally published.

On air charts, as on marine charts, prominence is given to aids to navigation, dangers, routes, and limits.

The main dangers are high ground and high structures; these may carry warning lights but are still invisible in cloud. Thus heights in feet above sea level are given greater prominence on air charts than on topographic maps. They are printed in bolder type and the maximum ground elevation in any sheet is enclosed in a frame for prominence. For structures like radio masts in the vicinity of an airport the height above local ground level may also be shown.

Hill shading is preferred to layer tints. One reason is that differences between various tints may not be apparent in the artificial light used in aircraft flying at night. This light is usually violet and hence only black and violet print on the chart are well seen. For the same reason, air information is printed in magenta.

Air information comprises airport and airfield symbols, lights, radio navigation stations, dangers, and limits of various flying control areas.

Topographical detail visible in daylight is printed in subdued colours and priority of choice is given to prominent objects identifiable from above such as lakes, major rivers, railroads, towns, and forests.

3.4 Thematic Maps

This class of maps, also variously referred to as topical, statistical, distribution, or special maps, has been and continues to be extensively discussed in other textbooks and cartographic periodicals, to which reference should be made. Organizations whose main purpose is to produce topographical mapping do sometimes engage in preparing special maps, for example a national atlas; a brief review of the basic features of such production may therefore be useful.

3.4.1 Source material

As with air information, the source material for special themes or topics has to be obtained from outside the map production organization. This involves consultation and possibly the formation of an editorial committee to decide on the best designs for the maps.

3.4.2 Projection

Statistical data such as density of population, land use, etc. should be presented on an equal-area basis. However if the area covered by the map is limited to one national territory, the difference between an equivalent projection and a conformal one is probably not sufficient to distort the statistics; existing topographical maps may therefore be modified to produce suitable base maps.

3.4.3 Sheet size and scale

In a national atlas it is clearly desirable to show the whole country on a single sheet. The equivalent paper size for several standard scales (1 : 1 000 000, 1 : 2 000 000, 1 : 2 500 000, etc.) can be computed to aid a decision on the paper size to be adopted. The larger the map the more clearly it will display the information. But a large atlas is expensive and difficult to handle. A3 (42 × 30 cm) is the most convenient of the standard paper sizes but may not be suitable for the shape of a country. It is quite likely that a non-standard paper size will be chosen.

3.4.4 Base map

Presentation of the thematic data must be the most prominent feature of each map, hence the base map must be both monochrome and subdued; a shade of grey is probably the best choice. Areas which on a topographical map would be filled with symbolic colour (lakes, forests, etc.), must be shown by outline only, or omitted, and relief is best shown by hachures. The base map can often be produced by deleting unwanted detail from the black, red, and blue plates of a small-scale topographical map and then combining these three plates into a single grey plate.

For economy, copies of the same base map should be used as a base for as many of the thematic maps as possible.

3.4.5 Distribution symbols

Portrayal of the geographical distribution of some topic may involve quantity only (e.g. rainfall, population) or quality only (e.g. type of vegetation) or both quantity and quality (e.g. population by ethnic group).

To illustrate quantity, point, line, or area symbols may be employed; the choice is often determined by the basis of measurement of the quantities to be shown. For example, rainfall is measured at a network of stations which are points. This is analogous to the measurement of spot heights. Just as contours or form lines are interpolated between spot heights and drawn as continuous lines, so *isohyets* (lines of equal rainfall) can be drawn between rain gauge stations. Further, just as layer tints are added between contours to show up low and high ground, so tints can be added between isohyets to show areas of low and high rainfall. By convention, rainfall tints range from brown or yellow in dry areas, perhaps through green to pale blue and dark blue for heavy rainfall.

Cities and towns may be regarded as points and their populations must be shown (on a small-scale map) by point symbols. A common method is to use solid circles with areas proportional to the population of each place. Rural areas are divided into small enumeration units and the total population in each unit is counted. If the boundaries of the units are shown on the map then the population of each unit can be shown in the same way as for a town, by drawing a circle of suitable size in the centre of the area. A better method is to use dots of uniform size, each dot representing, say, 100 persons. Then six dots in a unit indicate a population of 600. If the dots are evenly spaced, the unit boundaries can be omitted.

An alternative to showing total population distribution is to show density of population; the number of people per square kilometre in each unit is computed and suitable tints are added. A light tint indicates sparse distribution, a dark tint is for dense population.

To illustrate quality as well as quantity, suppose the population is comprised of five ethnic groups. Then each group is to be shown by a colour. One method would be to draw five circles of appropriate sizes and colours in each unit. Another method is to draw a *pie-graph* in each unit. This is a circle of size proportional to the total population, divided into coloured sectors, the size of each sector being proportional to the number in the group it represents.

Instead of using symbols it is of course possible to write a figure representing the number of inhabitants in each unit. As with spot heights, this gives no quick visual image of either distribution or density.

A topic such as 'Distribution of educational facilities' can be illustrated by designing a set of symbols to represent infant schools, primary schools, secondary schools, technical colleges, etc. Each symbol is then drawn in each unit at a size proportional to the number of schools of that class in that unit.

Representation of topics such as soils or crops or diseases is usually more complex because these are measured in terms of the areas they cover. If an area contains several different types of soil but the exact limits of each have not been surveyed one method of illustrating the position is called 'interdigitation': a set of parallel strips or fingers interlocking, each strip symbolizing one type of soil.

For the layman, pictorial or associative symbols are easier to comprehend. Distribution of livestock can be shown better by symbols representing cattle, sheep, goats, poultry, etc. than by geometrical symbols. But this treatment cannot be applied to a subject like minerals, for which geometrical symbols may be used or, probably better, initial letters or abbreviations, e.g. Bx = bauxite, Gr = Graphite, etc.

Topics involving movement can be illustrated by line-flow symbols. A simple example is the volume of traffic along a road or railway, which can be shown by a line symbol having a width proportional to the volume of traffic.

Time is a variable which is not easily illustrated on maps, the preferred method being by graphs or diagrams in which one axis represents time. For example it may be desired to show rainfall in each month of the year. One solution is to draw a graph for each rain gauge and place a miniature of the graph on the map at each gauging station. A better solution is to draw a set of twelve rainfall maps (one for each month) with the usual isohyets and tints and print them together at a reduced scale on one page.

As previously stated, this aspect of mapping is of great interest to many cartographers and gives wide scope for imagination and experiment.

Thematic maps should always be annotated with the source and date of the topical information. When a collection of thematic maps is put together as in a national atlas, they should be supplemented by a topographical map of the country, possibly at 1 : 1 000 000 cut up into suitable pages, and by topographical maps of the principal towns. Selected copies of old maps are always of interest, and no atlas is complete without the addition of a gazetteer listing all places named on the various maps.

3.5 Diagrams

A map production organization frequently has to produce diagrams on a map base both for internal and external use. There is a need for progress or index diagrams showing areas covered by triangulation, levelling, air photos of various dates and scales, mapping at various scales, etc. These may be in daily use and will also appear in annual reports and other publications. Diagrams of map series showing sheet numbers and sheet lines are needed for a map catalogue (see Chapter 8).

Such diagrams are virtually a form of thematic mapping. In fact the base maps used in a national atlas are usually equally suitable as base maps for diagrams of medium size.

Diagrams and maps intended to illustrate reports or to be included in any kind of book should preferably be of the same paper size as the book page (or capable of reduction to that size by a simple fold) since folded maps create difficulties in both binding and use and may get torn or detached.

4

In The Cartographic Drawing Office

4.1 Equipment

In a traditional drawing office the heavy furniture consists of large tables on which maps may be laid for inspection and other purposes, drawing tables having adjustable height and tilt, light tables having glass tops over a tubular light for work on transparent media, and cabinets of large drawers for storing uncompleted work and other materials in sheet form.

In or adjacent to the drawing office there should be one or more types of optical projector (optical pantograph) for enlarging (up to $\times$ 4) and reducing and direct transfer of revision details from air photos, and a coordinatograph (see Section 4.5).

More portable items should include a pantograph for manual enlargement and reduction of drawings, a micrometer microscope for measuring very small dimensions such as line widths, hand magnifying glasses, punch register equipment for punching holes or slots in drawing materials, register studs (see Section 4.7), scribing equipment (see Section 4.3), various equipment for lettering ranging from simple stencils to phototypesetting machines (see Section 2.6.5), and waxing equipment for mounting the names.

There must be an assortment of straight edges and standard steel scales, parallel rulers, a beam compass for long-radius arcs, curves and splines for drawing long curves.

Each cartographer should have his personal set of drawing instruments: 30 cm standard scales, set squares, protractor, dividers, compasses of various radii, eraser, knife, brush, and technical pens (with tubular points of various standard gauges from 0.1 mm upwards; there is, however, often an ink-flow problem with the finer points). Ruling pens are still in use but it is difficult to set them to give the precise line-width required.

For work on paper, standard spirit-based waterproof inks are used; but since most work is now done on plastic sheets, special plastic-solvent inks are largely required. The most important qualities for inks are that they should flow freely and be fadeproof. Since the colours on the final printed map are acquired from printing inks at the printing stage of production, the ink drawings for all colours are done in black; hence coloured inks are little used in the drawing office. Opaquing fluid is required for touching up negatives and scribed work and preparing masks.

Pencils are in frequent use and the most useful grades of hardness are 2H and 4H. Sandpaper, emery paper, or other sharpener is useful not only for pencils but for keeping fine points or edges on other instruments.

Since new designs of equipment appear on the market quite frequently, the best source of up-to-date details of items mentioned above is from manufacturers' pamphlets and sales-promotion literature.

4.2 Drawing Materials

There is an ever-increasing variety of drafting materials: paper and plastic of varying thickness and opacity and other qualities, and with various coatings.

The main requirements for a drafting material are that it should be: dimensionally stable, receptive to inks, strong, flexible, and either transparent or opaque according to the type of work for which it is required.

The most suitable material may also be the most expensive but it must be remembered that the published map is the final and visible product of a series of expensive operations (air survey, ground survey, computation, photogrammetry) and a relatively insignificant saving on drawing and printing materials resulting in an inaccurate or shoddy-looking map is nothing but 'spoiling the ship for a ha'p'orth of tar.'

The qualities listed above are necessary for the following reasons.

4.2.1 Stability

For multicolour printing, each colour plate is prepared on a separate drawing. Obviously these separate drafts must fit exactly together at any time otherwise the colours in the final map will not be 'in register' — for example the red filling in a road will not exactly fit the space between the black edge casings. Therefore the drafting materials must not change size with variations in temperature and humidity and passage of time.

4.2.2 Receptivity

The surface must accept ink so that it adheres and does not flake off when it dries. In addition, the ink must not spread, so that lines will retain their correct gauge.

4.2.3 Strength and flexibility

These qualities are related to thickness. The material must be thin for economy and good tracing quality but it must not wrinkle, tear, or crack. It should be able to pass through a copying machine with a roller mechanism without becoming distorted.

4.2.4 Transparency/opacity

Most operations in the drawing and printing stages of map production require that the material shall be transparent to white light and ultraviolet rays; however, some stages require that the map image shall be on opaque material.

In addition to paper, glass and metal have been used as bases for drafting maps, but none of these materials has all the required qualities. Paper is not stable, waterproof, or strong but it has so many useful qualities, particularly cheapness, that it will continue to be used for many purposes.

4.2.5 Plastic sheets

For precision work paper has been superseded by two forms of plastic sheet: polyvinyls and polyesters.

Polyvinyl is marketed under various trade-names of which the best known in map production is probably Astrafoil. It is made in thicknesses between 0.125 and 0.5 mm, in rolls and sheets up to 300 × 135 cm, transparent or white opaque, with one or both sides having a matt surface (which, unlike a polished surface, will accept pencil, ink, dye, and solvents). Its main disadvantage is that it is somewhat inflexible and brittle and may crack or shatter if dropped on edge.

Polyester is also marketed under many names (Melinex, Mylar, Durafilm, Kodatrace, Permatrace, etc.). It is more flexible and tough, can be folded repeatedly without breaking, and is made in thicknesses from 0.05 to 0.20 mm. Its main disadvantage is that it is non-absorbent; even special inks do not readily adhere to it. However the opaque coating for scribing (see below) adheres satisfactorily.

To overcome the disadvantages of both plastics, super-Astrafoil has a polyvinyl surface on a polyester base. This of course raises the price four or five times above that of untreated polyester.

4.3 Scribing

Drawing with ink produces a *positive* image, i.e. black or opaque detail on a white or transparent background. The technique of *scribing* produces a *negative* image, i.e. a transparent image on an opaque background. Whether a positive or a negative is preferable depends on the production process being used but, apart from this, scribing has certain merits compared with ink drawing.

The base material for scribing is usually a polyester transparent sheet coated with a thin layer or film of material which is translucent to part of the visible spectrum (usually red, orange, or green) but opaque to actinic light which is mainly concentrated in the blue-violet end of the spectrum. The action of scribing is cutting away the coating by the use of hard, sharp, pointed, or bladed instruments. It is claimed for scribing that lines produced in this way have a more uniform width than those drawn with a technical pen on a good quality white opaque surface, and also that a trainee can become competent in scribing more quickly than in pen and ink work.

For fine lines 0.025 to 0.2 mm wide, the scribing point is rounded and can travel in any direction, for lines 0.2 to 1.25 mm gauge the cutter is a chisel-blade and must move in a direction at 90° to the line of the blade-edge to maintain a constant width. There are special cutters for drawing two or more parallel lines, or dotted lines, etc. To retain precise size scribing blades are made from sapphire or tungsten carbide.

The cutter can be mounted like a pen-nib in a simple holder and used freehand for scribing small amounts of complex detail, but this mounting is not suitable for extensive line work, for which there are holders of varying complexity and cost. They have two, three, or four legs mounted on ball feet which can slide over the coating. Some models have a spring to regulate the pressure on the cutter and a lens to give a better view of the detail being scribed. A loose swivel holder will keep a chisel cutter blade at 90° to its line of travel.

Pecked lines can be scribed directly or can be made by scribing a continuous line and then intermittently deleting it by application of opaquing fluid at regular intervals. By either method the pecks are likely to vary in length. One alternative method is to stick down over the scribed line a strip of preprinted transparent material ressembling a ladder, the black rungs of which blot out the scribed line at regular intervals. This is excellent on straight lines but more difficult on complex curves.

The scribed sheet is a negative image; as with an ordinary photographic negative, positive prints can be made by contact exposure to bromide paper. The scribed sheet may itself be converted into a transparent positive by adding a special black dye which adheres to the bare base exposed under the scribed detail; the unscribed coating is then removed by a special solvent. This process works well when the base is polyvinyl, but the dye does not adhere readily to polyester.

If the scribecoat is yellow or yellow-orange then a positive can be made by contact copying on to *autopositive* (or *direct positive* or *autoreversal*) film by exposure first to yellow light then to white light. This technique can be used also to make a combined positive from the scribed detail and black letter-press (or other detail) mounted on the scribecoat. The first exposure to yellow light is developed and produces an image of the opaque black work only; then exposure to white light makes a latent image of the scribed work only. Final development reveals a combined image (see also Section 5.4).

Extensive line work such as a contour plate is most efficiently prepared by scribing, but best results for small complex detail are still obtained by pen and ink.

4.4 Plotting the Grid

The position of topographical detail is plotted photogrammetrically (and sometimes on the ground) in relation to control points whose positions have been accurately identified, measured, and computed. These positions may be stated as latitudes and longitudes but most commonly are coordinates measured on the national grid. Therefore the first task in preparing a map is to put down an accurate grid in relation to which all the material to be used in compiling the map can be accurately positioned.

Grids have been discussed in Section 1.8 where it was indicated that for topographical maps on scales of 1 : 500 000 or larger a grid with spacing between the lines of 1 to 10 cm is desirable.

The simplest way of producing any required grid is to keep a set of master grids accurately drawn with lines at the various standard spacings and printed on transparent plastic. These master sheets must be larger by about two grid squares in each direction than any map likely to be produced. The master with the desired spacing is selected and a contact copy is made on transparent plastic. Suitable numbers are given to the grid lines, the sheet corners are plotted and sheet edges drawn and the unwanted ends of the grid lines outside the sheet lines are deleted by any suitable method.

An alternative is to keep a metal template with fine holes drilled precisely at 1 cm intervals. Coloured lines on the metal at two, four and five cm intervals indicate which holes are to be used for grids at these intervals. A special pricker placed in any hole in the template will prick a fine mark in the underlying plastic sheet at precise centres to give the required grid intersections.

If the grid is for a map of some non-standard scale with an interval other than one of the standard intervals, then it is best drawn by coordinatograph.

4.5 The Coordinatograph

This is an instrument designed for the rapid and accurate plotting of (x,y) or (N,E) rectangular coordinates: see fig. 4.1.

In the usual design there is on the left side a 'y' graduated track along which runs a gantry carriage. Attached to the carriage at right angles to the 'y' track is the 'x' graduated beam along which another carriage carries the plotting head. The track and beam are precisely graduated in millimetres. The carriages have clamps, slow-motion micrometers (reading to 0.01 mm) and magnifiers so that they can be set to precise values on the scales. Moveable tapes printed with values for various scale ratios lie alongside the precisely engraved scales and act as indexes to the graduations. As the tapes are moveable, the zero or other value can be placed at any desired point.

If the y (or x) carriage is clamped at a chosen value and the other carriage freely moved, the plotting point will draw an E–W (or N–S) line at the value fixed on the y (or x) scale.

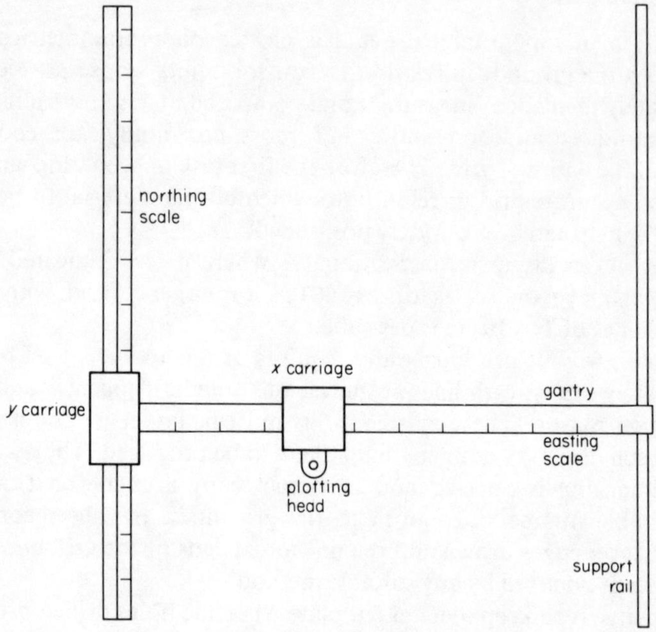

Fig. 4.1 Coordinatograph

The plotting point may be a pencil, pen, scribe-cutter, or light spot; the last is used for plotting in a dark room on to photographic film — see Section 7.13.

Coordinatographs range from small, relatively simple instruments to electrically-powered high-speed equipment. Under manual control the latter type has an illuminated indicator which displays the coordinates of the position of the plotting point. Other models are controlled automatically by computer (see Section 7.12). Such an instrument can be programmed to draw, for example, the curved lines of a graticule, which is otherwise a tedious manual process, especially when the radius of a curve is large or varying.

It is more usual to draw the graticule by taking out the grid values of selected meridian/parallel intersections from volumes of tables (of which the most useful are those published by the US Army Map Service for the UTM projection). These intersection points are then plotted by coordinatograph and joined by straight lines (at medium scales) or curves (at small scales) drawn with the aid of splines.

For graticule intersections on a projection for which there are no published tables, the geodetic section of a map production organization may be asked to supply values.

Coordinate plotting is completed by plotting selected ground survey control points. These will be points having permanent marks on the ground and may appear on photogrammetric machine plots and existing published maps. They are both an aid to and a check on the accurate positioning of the material from which the map is to be compiled.

4.6 Compilation Material

The material from which the map is compiled varies with the type of map to be produced: basic, derived, or thematic (see Section 1.10 for definitions).

For basic mapping the main material is photogrammetric machine plots made from recent air photos of the mapping area. Some mapping organizations have their own aircraft, air camera, and photogrammetric plotting equipment while others obtain the required photos and/or plots from commercial contractors.

The plots are on transparent plastic and show all the main ground features visible from the camera at the time of photography. They cannot show invisible detail such as names, descriptions of buildings, types of road surface, small objects, detail obscured by trees or shadows, undemarcated boundaries, etc. Such information has to be collected on the ground but this is usually done at a later stage in production (see Section 4.10).

Orthophotography

This is a possible alternative to machine plots. Air-photo prints are perspective projections of the ground surface, and detail appears displaced from its true plan position due both to tilt of the camera axis and to variations in surface elevation. Orthophotographic equipment is expensive but it can automatically rectify the displacements to produce a photograph in which the position of all detail is planimetrically accurate. However, to convert orthophotos into a line map requires skill in photo interpretation which the average cartographer does not possess. Annotated air photos may be used as a final map document in their own right out rarely as a stage in standard map production.

The orientation, size, and shape of machine plots (and orthophotos) is determined by the direction of flight lines, scale of photography, and extent of overlaps between successive photographs. The cartographer would obviously like to have complete information (including the invisibles mentioned above) on all detail to be shown in the map available at the start of fair drafting, but in fact it is found to be more practical to compile the draft map from the various machine plots and then send out a copy for *field completion* (Section 4.10). Machine plots are of course in monochrome (usually pencil) and are plotted at the required drafting scale (see Section 4.9).

For *derived mapping* the compilation material normally consists of copies of latest editions of basic maps covering the area. For example, a 1 : 250 000 sheet covering 1° × 1½° may be compiled from six sheets each ½° square at 1 : 100 000 or 24 sheets each ¼° square at 1 : 50 000. The larger-scale maps are accurately reduced to the required drafting scale by a process camera (see Section 5.3). If a 1 : 50 000 scale map is thus shrunk to 1 : 250 000 areas are reduced to 1/25 of their original size and much of the map detail may become unreadable. Hence it may be necessary to reduce in two stages, making preliminary compilation at an intermediate scale.

Although the original map may be in several colours, the reduced camera print will be in monochrome. Since the fair drawing for each colour of the derived map is drafted separately, there are obvious advantages in getting separate reductions for each colour of the basic map. This is no problem if the original drawings for the basic map are still available. If there is only a full-colour printed copy then colour-separation photography may be used to produce separate reductions for each colour. In dealing with colours separately the normal problems of exact fit (register) between different colour plates must be kept in mind.

4.7 Register Marks

Any one copy of the final map delivered by the printing machine contains all the information designed to be shown on it, but in preceding stages of production that information was probably drafted on several different sheets (*flaps* in American usage). Even a monochrome map may be built up from a standard margin (see Section 2.7.1) on one flap, a grid on another, and the topographical or other detail on a third. Each additional colour requires at least one more flap.

Obviously these separate flaps must all fit exactly over each other. A common practice is to use the four corners of the neat line (Section 2.7) as register marks, tracing them from the standard margin sheet on to each of the other flaps before drafting starts on the latter. A much superior method is to punch holes or slots in each piece of drafting material before the work starts. This enables flaps to be superimposed with greater accuracy and speed at any time, even in a darkroom. There are various possible arrangements (see fig. 4.2). Holes or slots in the west and north margins cause least hindrance to the draftsman. There should be in each of these margins either at least two holes or two slots or one hole and one slot, the spacing depending on sheet size.

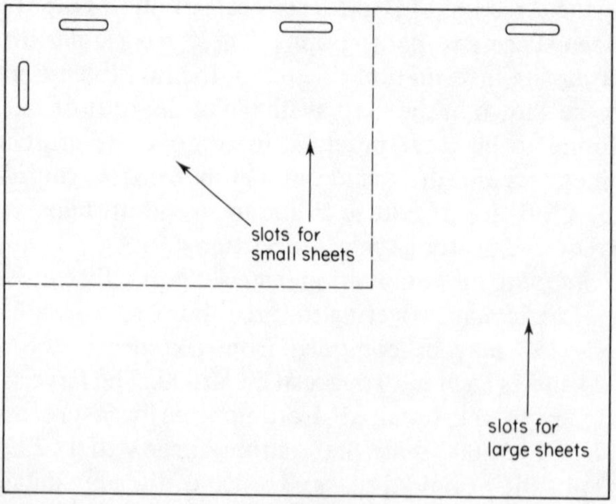

Fig. 4.2 Register slots

Punching is done with a special punch-register device. The various flaps can be accurately superimposed at any time by inserting register studs in the corresponding holes or slots.

4.8 Colour Plates

To produce a multicoloured map a separate printing plate is required for each colour of ink, hence a separate drawing must be made for each colour. Black may be regarded as the master colour since the black place usually carries the sheet edges, the grid and graticule, control points and other accurately fixed points of detail. The drawing of this takes priority and all other colours are then keyed to it (or copies of it, if other colours are drawn by different draftsmen).

For flaps to be produced by ink drawing the compilation material is assembled (over a light table if the material is transparent) in correct position under a transparent plastic sheet and the required detail for that particular flap can be traced off.

If the flap is to be scribed (e.g. a contour plate to be printed in brown) then it is convenient to have the compilation detail (key) printed on the scribecoat surface. A light-sensitive coating is added to the scribing surface (if this has not already been done during manufacture) and the compilation image transferred to it by contact exposure. Then the contour lines are scribed through the image lines. If the production process calls for a reversed (wrong-reading) scribed flap, then the compilation can be printed down in reverse. Many draftsmen prefer to scribe on a *negative* key, i.e. on transparent detail with an opaque background.

4.9 Drafting Scale

When man-hours cost less and all drafting was done by pen and ink, maps were often drawn at larger scales (up to double) than the intended publication scale. The finished drawings were reduced by camera to the printing scale.

This technique reduced the size of imperfections in drawing symbols, line widths, and small lettering but greatly increased the total volume of work to be done. There were also the inevitable problems of register between the various reduced plates.

New techniques such as machine-set lettering, scribed lines, and preprinted symbols have made large-scale drafting unnecessary and normal procedure now is to draft at publication scale.

4.10 Field Completion

This is a stage in production applicable only to basic mapping. It is a specialized field task usually done by a surveyor, but is a useful experience for a cartographer. The purpose is to add details which cannot be obtained from air photos or office sources; recent developments since the date of the air photos may also be picked up. The operation can also be used as a check on the work done in the drawing office.

When all available point and line detail has been drawn on the separate colour plates, a combined print may be made from these plates. Normally a combination of the detail (black and red) and water (blue) plates will suffice; vegetation (green) may sometimes be included. A separate transparent print of the contours may be made but should be omitted from the combined print as it has too much line work which may confuse with other line detail. The combined print is monochrome, which permits the field information to be added in various colours.

The surveyor will also take a set of air-photo prints and any existing maps (at any scale) of the area, including adjoining areas.

The information to be collected varies slightly according to local circumstances. The surveyor enters it on the print according to a colour and symbol code (which is not the same as the one to be used in the published map).

New detail: any features changed since date of air photos.
Gaps: add detail missing in areas obscured on photos by shadows or trees.
Roads: describe surface, describe crossings of other line detail (bridge, ford, levelcrossing, etc.), add kilometre posts, show destination at sheet edge.
Tracks: indicate principal motorable tracks, including remarks like '4 wheel drive only'.
Airfields: add type of runway surface.
Pipes and overhead lines: show turning points and points of crossing roads and railways.
Buildings: describe isolated buildings, and pick out important buildings (e.g. hospital) in built-up areas.
Mines, pits and quarries: identify product.
Water: add springs, wells, boreholes, pumps.
Plantations: identify crop.
Names: collect and list all place names (see Section 2.5).

4.11 Office Checks

On receipt of the field completion sheets, the fair drawings and scribe sheets can be amended in the cartographic office. Area tint symbols may then be added (see Section 5.7). Then, before the flaps are sent for colour proofs, they are subjected to a final check. This is done flap by flap starting with the black plate.

On all flaps the following must be checked:
(*a*) The material must be clean (spots of dirt will reproduce).
(*b*) Positive line work and letterpress must be opaque. Holes and pale lines will print badly.

(*c*) Letterpress lines must run parallel to the north and south margins (unless intentionally oblique, as along a river).
(*d*) Correct registration of all colours with black (the master colour).
(*e*) Comparison with adjoining sheets: straight features across a sheet edge must not bend at the edge; features falling partly in each sheet should have the same name on both sheets.
(*f*) All symbols, line gauges, type sizes, etc. to be in accordance with specification.

Black plate: check dimensions of grid, graticule and sheet corners, check all internal detail and all variable marginal information.

Brown plate: check spacing of contours, fit of contours with rivers, numbering of contours, values of contours consistent with spot heights.

All plates: check that work on one plate does not obscure or clash with work on another. It may be necessary to break some lines, clear some area tints, move some names etc.

Checking must be organized on a methodical basis. For line work check along a line, e.g. a railway; for point detail it is best to examine the map grid square by square.

4.12 Classified (Security) Information

There are some topographical details which are either omitted or modified on maps, generally for reasons of military security. A non-military mapping organization has to seek directions from the appropriate authorities. To take an example without military implications: it is considered undesirable to show any detail within a prison as this might be an aid to planning a gaol-break. At most, only the outer perimeter (which is anyway visible to passers-by) should be mapped.

In these days of high-altitude aircraft and orbiting satellites equipped with long-focus cameras and other sensors, there is little that can be hidden, even when camouflaged; in fact blank areas on a map only draw attention to the fact that something has been omitted, and excite curiosity in interested quarters.

Apart from omitting interior details, as in a prison, 'give-away' descriptions such as explosives factory, gun-site, radar station, should obviously be omitted, although many such establishments identify themselves on the ground either by their appearance or even by notice boards proclaiming their business. Hence blanks on maps do not really conceal information but make it slightly more difficult to obtain. There seems to be no point at all in omitting or not describing such obvious features as airfields, power stations, radio stations, although some security authorities would have all these deleted from maps.

4.13 Production Planning Control and Records

A small map production organization can probably work reasonably efficiently on an *ad hoc* basis with day-to-day decisions by one or two executives on what is to be done and who is to do it. Very large organizations need managers trained in work-study, critical path techniques, organization and methods, business efficiency, etc., with the general objectives of producing maps by methods in which quality, speed, and cost are properly balanced.

Each organization will devise its own system, covering all departments. In a large cartographic office with many map sheets in work at the same time, a wall progress chart is useful for showing proposed dates (in pencil) of various stages of work and actual dates (in ink). Such a chart aids planning, control, and allocation of work to the various technicians and pieces of equipment.

Each map sheet (including each sheet of a series) should have a file, dossier, or history sheet in which are recorded production instructions, reference to specifications, compilation material, techniques (to be) used at each stage, names of staff engaged and hours spent, relevant dates, number of copies to be printed. The file is brought up-to-date at each reprint or revision of the sheet, at which times it provides valuable information.

5

Reprography

5.1 Transfer of Images

The later stages of map production are processes for converting the drawings made in the cartographic office into multiple copies on paper, which are the published maps, plans, or charts. There is a variety of reprographic (copying or printing) techniques and as with any technology, explanation is easier if the layman first becomes familiar with the technical terms which recur most frequently.

Copying (from the same Latin root as *copious* = plenty) literally means making large numbers of copies; technically it means making directly from the original. *Printing* literally means making copies by use of pressure; technically it strictly means that the copies are made from an image on a metal plate. However the two words are often loosely used as if they had the same meaning. To increase the confusion, 'copy' is also used in 'the trade' to mean the original which is to be copied.

Plate is used to mean not only the metal printing plate used to transfer ink images to paper but also is loosely applied to drawings and their images at various stages in the production line (*flaps* in American usage).

The images at these various stages may be *right-reading* or *wrong-reading*. The latter is a mirror-image of the former. A right-reading image on a transparent medium (e.g. a tracing) will appear wrong-reading if the tracing is turned over and viewed from the back. 'Reversed' is sometimes used to mean wrong-reading but this can lead to confusion as this word occurs in other contexts such as tonal reversal.

Images or copies may also be *positive* or *negative*. In a positive copy, relative tone or lightness/darkness values are the same as in the original: a black line on a white or transparent background will still be a black line on the copy. In a negative copy the tonal values are reversed: the black line becomes white or transparent on a black or opaque background.

Historically, all copies of documents were perforce made by hand one at a time until the invention of printing, which in Europe was five centuries ago. Even then, for a further four centuries the image on the printing plate had to be made directly by hand. Rapid and precise image processing had to wait for the discovery of light-sensitive chemicals and the invention of photography.

5.2 Process Photography

The main groups of light-sensitive chemicals which are used in map production processes are silver halides, dichromates, and diazo compounds. The two latter have comparatively slow reaction to light and produce positive copies from positive originals; silver salts react rapidly and normally result in tonal reversal, i.e. they produce negative copies from positive originals and vice versa.

The silver halide is prepared in the dark by adding silver nitrate to a suitable halide, usually a bromide, in which case the resulting reaction gives silver bromide. Since this is non-adhesive it is mixed into a suitable base, usually gelatine, to produce a transparent emulsion which can be applied to glass plates, film, plastic, or paper and dried to give a light-sensitive coating.

On exposure to light the bromide is chemically changed, the amount of change at any point being proportional to the amount of light falling on that point. At this stage there is a latent image which is not yet visible. A developer is applied which reduces the exposed bromide to black particles of colloidal silver. The emulsion is then fixed by adding hypo or cyanide which makes the unexposed bromide soluble in water and it is then washed away. The plate is now no longer sensitive to light so that it can be dried and inspected. The image is in negative form and will have to go through a similar process to produce a positive copy.

The process can be effected through a lens (i.e. by a camera) in which case the original should be an opaque material and well illuminated from the front; the lens can be manipulated to produce an enlarged or reduced image. Or it can be by contact, in which case the original must be on transparent material, lit from behind. In either case, the resulting image is wrong-reading. In everyday domestic and commercial photography the first stage is through the camera lens, giving a wrong-reading transparent negative. From this can be made a contact print of the same size, or by exposure through another lens to obtain an enlargement. Either way the end result is a right-reading positive which is usually on paper.

The simplest silver halide compounds are used to prepare *process* or commercial film and plates. These are sensitive only to the shorter wavelengths in the spectrum: blue/violet/ultraviolet (350–500 nm). This has two advantages in map production: (1) the plates are insensitive to red light so the latter can be used to see in the darkroom during processing; (2) blue appears to the camera the same as white; therefore it is possible to print a compilation as a blue key on a white drawing surface, ink the required detail in black, and photograph it, when only the black will appear on the camera plate.

By adding various dyes to the silver bromide it can be made additionally sensitive to green light (orthochromatic film) or to all visible light (panchromatic film). These are only needed if it is necessary to photograph a coloured map.

5.3 The Process Camera

Same-scale copies of a monochrome drawing are made by a contact process (see Section 5.4). Colour copies of a multicolour original can be made by camera using colour film and colour print paper but this is a highly uneconomical way of producing more than a few copies of an existing multi-colour map.

The main function of a process camera is to make accurate copies at scales different from that of the original, particularly reductions. For example if a 1 : 200 000 scale map is to be compiled from four sheets of a 1 : 100 000 map, the camera is used to make exact half-size reductions of the latter.

Material to be photographed is preferably drawn on an opaque base, but transparent material can be backed by a white base to give good copy.

Process cameras are large instruments and may produce plates as large as 125 × 150 cm. A typical lens has a focal length (*f*) of 625 mm and can enlarge or reduce by a factor of 5 or more. Some cameras have two lenses, or inter-changeable lenses. Modern camera installations require two rooms. The camera is mounted in one room with its optical axis horizontal and its rear projecting into another room, the darkroom, where the light-sensitive plates can be safely handled. The lens in the open room points toward the vertical copyboard which has a vacuum frame and is mounted on a carriage so that its distance from the lens can be adjusted (see fig. 5.1).

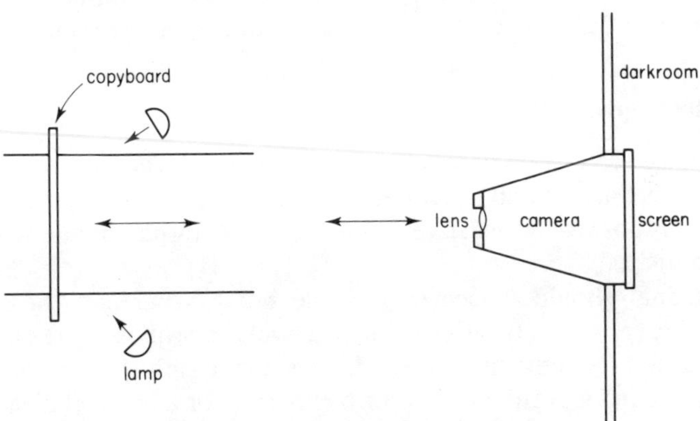

Fig. 5.1 Dark room process camera

The original material is mounted on the copyboard, held in place by a glass cover, flattened by a vacuum pump, and brightly illuminated by suitable lamps. For photographing coloured work on to panchromatic emulsion, a white light is necessary and pulsed xenon is a powerful source; tungsten or quartz halogen lamps may also be used. For copying black detail on to orthochromatic emulsion, the light must be strong in the blue–violet–ultraviolet range for which suitable sources are metal halide, mercury vapour, or carbon arc. However, the latter is no longer considered compatible with modern standards of health and safety.

The image of the original 'copy' created by the lens falls initially on to a ground-glass screen at the rear of the camera where it is viewed by the operator in the darkroom. The distances of the lens from the copyboard (u) and from the screen (v) are automatically adjusted to give correct magnification (v/u) and focus ($1/f = 1/v + 1/u$). The lens shutter is then closed, the screen is replaced by a photoplate or film and the exposure is made.

In addition to their main function of accurate scale changing, process cameras are used for several other cartographic tasks such as:

Increase of contrast by use of special litho film:
Colour separation by use of filters;
Rectification of maps or flaps which have shrunk or are otherwise distorted;
Thickening or thinning of lines by placing a device in front of the lens or by varying exposure time.

Other types of camera are used in the photostat (see Section 5.17.1) and microfilming (Section 8.6).

There is a present tendency to use a microfilm camera instead of a process camera for many jobs because it is quicker and there is a large saving on negative costs. For example, to reduce a 1 : 50 000 map to 1 : 250 000 the microfilm camera may reduce the original by 17.5 to 1 on to a 35 mm negative which is then enlarged × 3.5 in an enlarger to give a final copy at the required scale.

5.4 Contact Copying

To produce same-scale copies by a photographic process, a camera is not required. There are several processes and the choice is determined by whether the original is positive or negative and opaque or transparent and what type of copy is wanted.

The original should whenever possible be transparent. Then to get a reversed-tone copy (e.g. positive from negative) the print is made on to a silver halide emulsion (or ferro-prussiate, which is less sensitive); to get a same-tone copy (positive from positive) diazo or bichromate or a reversal bleach process can be used.

If the original is on an opaque base, a transparent copy may be made by a 'direct positive' or reflex process, using yellow light. The somewhat complex photochemistry of this makes use of the Herschel effect.

If ortho film is exposed to blue light and developed, it will produce a black negative. If it is not developed the photochemical change remains latent. If it is exposed to a light to which it is not sensitive, e.g. yellow, the latent state is cancelled, and development would then produce clear film. Direct positive film is exposed to blue light at the factory. If it is later exposed under yellow light in contact with a positive tracing of a map flap, the non-image areas will develop clear and the image areas black, i.e. a positive print from a positive original.

To get a sharp image on the copy, the original and the sensitized copy surface must be in close contact. Copying of large drawings is therefore carried out in a vacuum frame, which is closed by a glass cover and exhausted by an airpump, causing the surfaces to be pressed tightly together. The original must be laid face down on the copy material. If the drawing is face up the light will diffract or diffuse into the transparent material under the opaque details on the original; even when the material is quite thin such diffusion can spoil the sharpness of finely-drawn work.

From such a face-down contact, the resulting copy will be wrong-reading. This must be kept in mind but in practice it is rarely a disadvantage because there is frequently a later stage in processing which requires further contact copying, at which stage the wrong-reading image will reverse back to right-reading.

Copying may be on to any material coated with a light-sensitive emulsion. For example, from an air photo film negative, positive prints are made on to opaque white paper, or diapositives are made on glass or plastic for use in plotting machines.

Timing is a very important factor in all photochemical work, both in exposure through a lens and contact copying. Too short an exposure produces a weak image; for example, lines appear thinner, whereas over-exposure produces a coarse image with lines appearing thicker.

This effect can sometimes be useful in map production. If, say, a map at 1 : 63 360 is to be republished at 1 : 50 000 without redrawing, the scale can be changed by photographing through the camera, and deliberate under-exposure (or under-development of the image) may be used to prevent equivalent increase in the width of lines.

5.5 Colour Production

In the foregoing, particularly in the sections on symbolization, there have been many references to colour. At this point in the production line it is necessary to consider the theory and practice of colour production in some detail.

In general, the colour of simple point and narrow line symbols on the published map will be the unmodified colour of the printing ink used to print those symbols and this colour can be chosen and ordered from an ink maker's chart. To obtain the required colour for areas (and also for points and lines if the four-colour process is being used — see 5.8) is a more complex exercise.

Colour is a sensation caused in the average human optic nerves and brain when certain wavelengths of light fall on the eye. This is not a very satisfactory definition because there is no standard human optical system. If a set of nerves tuned to receive a particular colour band does not function properly then the total colour sensation will be different from that of an eye/brain with normal receptivity; this is called colour blindness. It is not a rare condition and there might be quite a good market for colour maps produced specially for colour-blind people.

A very large number of different colours can be distinguished in the visible spectrum produced by a rainbow or prism, ranging from red to violet. They can, however, all be obtained by mixing in various proportions the three *primary* light colours, which are red, green, and blue. This is called *additive mixing*. Correct proportions should give white light.

The technical terms most frequently appearing in discussions on colour theory are the following.

Chromatic: having hue.
Achromatic: without hue, e.g. black, grey, white, silver.
Hue: a chromatic colour.
Saturation or chroma: purity of colour. Saturation is reduced by adding any achromatic colour to a hue.
Lightness or value: the strength of light apparent in a colour. Of the pure spectral hues, yellow appears the lightest.
Tone or luminosity: can be measured against a standard grey scale ranging from black to white.
Tint: a mixture of a hue and white, giving greater brightness or value.
Shade: a mixture of a hue and black, giving decreased brightness.

Cartography is less concerned with additive light mixing than with *subtractive* effects. For example, a colour filter is only transparent to certain wavelengths and absorbs others. A blue filter absorbs all but blue (or possibly a range of wavelengths which when viewed in combination appear blue). If a multicoloured map is photographed through a red filter, only map detail which is red or contains red will form an image on the negative. This is a method used for making separate colour plates of a map when the original drawings are not available.

The most important colour phenomenon in map making (and in everyday life) is *reflection subtraction*, which governs the apparent colour of everything (except a light source) that we observe. White paper appears white because it reflects all wavelengths. Black ink appears black because it absorbs almost all light falling on it. Red paint appears red because it reflects most of the red wavelengths and absorbs most of the others. If illuminated with light containing no red (e.g. light passed through a non-red filter) it will reflect nothing and appear black.

Complementary to the three additive light primaries, there are three *subtractive primaries*: cyan (minus-red), magenta (minus-green), and yellow (minus-blue). On mixing two primary pigments, say cyan and magenta, the cyan absorbs all red light, the magenta absorbs all green, leaving only blue to be reflected. Hence a mixture of cyan and magenta appears blue. Addition to the mixture of yellow, which absorbs blue, will give an end-product which appears black.

There are several ways of manipulating filters and photography to get a desired result. Suppose that we have a map on which boundaries have been printed in magenta and we want to obtain a plate showing only the boundaries. Then we try photographing it through a green filter. Magenta contains no green and so no light from the boundaries should pass through the filter. Light from the white paper and all other colours will pass. On the developed camera negative, the image of the magenta lines should be transparent, the remainder of the negative opaque. Then a contact positive from the negative will give a flap of the boundaries.

5.6 Tints and Shades

It would be possible to obtain various tints of one colour on a map by diluting the coloured printing ink, thus reducing its saturation; similarly shades of a colour could be obtained by mixing with black ink. But since only one ink can be applied to any one printing plate, each such tint or shade would need its own plate, which would be highly uneconomical. Another method must be used to get all the tints of one colour on to one printing plate.

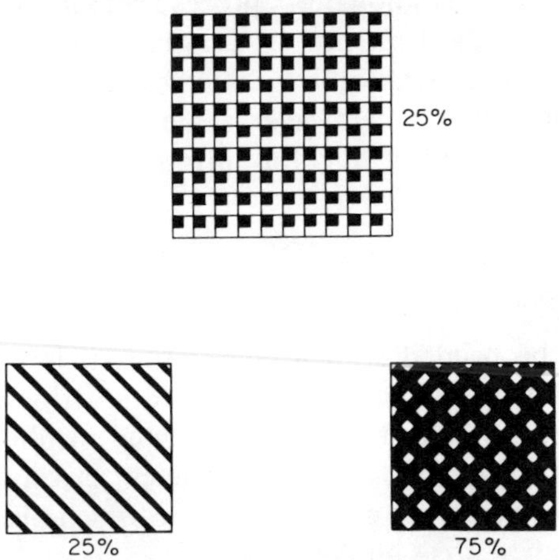

Fig. 5.2 Percentage screens (enlarged)

Percentage tints (see fig. 5.2)

Suppose the requirement is for a 25% tint (i.e. 25% full colour, 75% white). This can be obtained by a stipple or a ruling. Imagine the area divided into equal squares and in each (white) square place a dot of full colour, the dot size being a quarter of the square size. If the squares and dots are too small to be distinguished by the unaided eye the surface will appear as a continuous uniform tint of the required brightness. The dots (called a *stipple*) are usually circular but may be square or any other shape.

Instead of dots it is possible to use lines (which are more easily drawn). For a 25% tint the (white) spacing between the lines should be three times the width of a line. This is called a *ruling* and with very fine lines gives the same appearance as a stipple.

Two sets of lines at right angles will give a *cross-hatch* or cross-ruling. If the lines and spaces are the same width the result will be a 75% tint, which is difficult to distinguish from 100% full colour. However a contact negative could be made from it which would be a 25% dot stipple.

It should be noted that during the various processes from screen to printed map, the size of dots (and lines) tends to increase by up to 12%. This must be taken into account when constructing percentage screens (see Section 5.7).

Shades

Whereas tints only vary the brightness of a pigment shades change its apparent hue. For example, some shades of red or orange appear brown. Shades are made by adding a grey tint (e.g. a 25% dot stipple of black) on to a colour (hue) which may itself be a tint or solid.

By combining various percentage tints of the three subtractive primaries (cyan, magenta, yellow) and black, almost any desired colour can be produced. The technique is known as 'process colour' printing and has for many years been a standard method of printing area colours. For printing points and lines precise registration is vital which causes map producers to be wary of adopting the technique. However, some major producers now print all their maps by this method, using only four printing plates and inks (see Section 5.8).

5.7 Screens and Masks

If an area is to be printed with a colour tint, it is not necessary for a cartographer to fill that area on the flap with a dot stipple or line ruling; as mentioned in Section 2.2.4, sheets of preprinted symbols or tints (called *screens*) can be purchsed. They may be permanent or expendable. The latter are printed on thin transparent plastic foil from which pieces can be cut and mounted on the required areas on the drawing.

The lines or dots on tint screens are drawn at a large scale and photographically reduced so that they are spaced at 40 to 80 per centimetre. Screens may be obtained in sets covering a range of percentages, e.g. 10, 25, 50. There is no value in having many percentages in such a series because the difference between, say, 60% and 70% is barely perceptible. Screens may be supplied in positive or negative form.

Line ruling screens are easier to make and in the past were the most common; however, dot screens give better results and are now the most used.

Screen angles

If two or more screens are used over the same area (e.g. on the black plate and the yellow plate to produce a buff) care must be taken to avoid getting moiré effects, that is, undesirable patterns of light and dark lines. This is avoided if the screens are placed so that the line rulings or lines of dots will intersect at angles of 30° or 45°. Permanent screens are produced on somewhat thicker material with a sheet size of probably 50 × 60 cm. They have to be used with *masks*.

5.7.1 Masks

A mask is a transparent sheet of plastic which is laid over the drawing and has opaque paint applied to it over particular areas. The area to be opaqued need not exceed half the sheet area: if the total area to be printed with a tint is less than 50% of the sheet area, then a *positive mask* is produced by opaquing the tint area. This is converted to a negative mask by contact photocopying; the negative mask has a clear window over the tint area and is opaque elsewhere. If the area to be tinted covers more than half the map then a negative mask can be made directly by adding opaque over all the non-tint area.

5.7.2 Peel-coat

The use of opaque paint has to a great extent been superseded by the use of plastic sheets which have an opaque coating which can be peeled or stripped off. The boundary of the area to be stripped can be traced but a better method is to sensitize the surface (this may already have been done during manufacture) and print on it an image of the boundaries of the area to be tinted. The boundaries can be cut with a knife or by chemical etching; the coating over the tint area can then be lifted off.

We now have a negative mask which defines the area to be tinted and a choice of screens to provide the required density of tint. There are several possible procedures from this point. Probably the simplest one is to make a plate with the required areas tinted and no other detail. A photofilm is placed in the vacuum frame in the darkroom with a negative tint screen over it and the negative mask on top image-side down (wrong-reading as seen from above). After exposure and development there will be a positive tint in the required area (right-reading) and clear unexposed film elsewhere.

Combined tints

All the tints, plus point and line detail, which are to be printed with one colour of ink, must be combined into one plate. One mask and screen are exposed as above; then, instead of developing, masks and screens for other areas are successively exposed, as is a negative of the point and line detail. Finally the plate is developed and fixed.

5.7.3 Halftone, variable shading and vignettes

On an ordinary photograph, fully exposed silver bromide is all reduced to silver and the result appears black; areas exposed to less light and/or for shorter periods are not fully converted to black and appear in various shades of grey.

But on a lithographic printing plate, printing ink cannot be made to behave in the same way. As explained below, the ink either adheres to detail on the plate or it does not adhere where there is no detail; where it adheres it will produce (if black ink is used) a black image and elsewhere no image. Ink will not dilute itself to give shades of grey.

As explained above, greys (and other shades and tints) can be produced by the use of screens; but the screens described above will each produce only a uniform tint and not one varying in density from place to place.

To produce varying tone there is a special screen called a *halftone* screen. There are two principal designs: one used for the past century or so is made by a close ruling of fine parallel lines on a thin glass plate; the lines are opaqued and two such plates are placed with the lines in contact but an angle of 90° between the two sets. This leaves a network of transparent squares between the sets of opaque lines. A more recent design of screen has a close pattern of small dots on plastic; the opacity of the dots is graded from fully opaque at the centre to transparent at the edge.

Light passing through the traditional type of screen on to photographic film produces a set of discrete dots and the size of each dot is proportional to the amount of light received. The dots are very small so that the unaided eye sees them as continuous shades of grey. With the newer type of screen the light passes through the non-dot areas but the apparent result is the same. A small magnifier reveals the structure of the image. For printing copies of photographs, paintings, etc., in newspapers, magazines, and books this is the technique used.

The map maker only needs to use halftone screens if it is required to print hill shading (see Section 2.4.6) or vignetted boundaries (Sections 2.2.12 and 2.3) in both of which the strength of shade or tint varies rapidly over a short distance.

The *hill-shading* plate prepared in the drawing office is photographed through a halftone screen to obtain a negative from which the image can be transferred to a printing plate.

Vignetting is slightly more complex. Suppose it is required to vignette the edge of a lake, with solid blue colour at the shoreline fading out to white in deeper water. The negative is prepared as follows (see fig. 5.3).

In a contact printing frame place the unexposed photofilm, emulsion side up, and above this the halftone screen, then a positive mask (face down) of the land (which will have a window over the lake). Then follows the glass top of the printing frame but since this alone is not thick enough for the process, a further glass spacer or separator (which may be a centimetre or more in thickness) is placed on top. Above this comes a positive mask of the lake (opaque over the water) and finally the diffuser, which is a transparent plastic sheet.

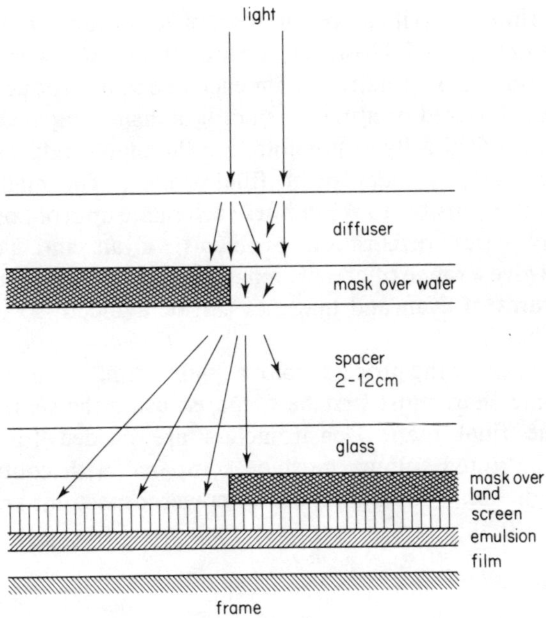

Fig. 5.3 Vignetting a shore-line

The whole assembly is placed under a suitable lamp and when this is switched on, the light falling on the diffuser from above starts to scatter so that some diverges from a vertical path to spread under the edge of the opaque water mask then through the spacer into the edges of the window in the land mask. The amount of light diverging from the vertical decreases as the angle of divergence increases, so that the amount reaching the half tone screen decreases away from the shore, the screen in turn transmits a decreasing dot size to the negative.

The width of the vignette band is controlled by the thickness of the separator, and the strength of colour in the band by the exposure time.

Double lines

A similar technique can be used to produce a double line (e.g. a road casing) from a drawing of a single line. Light passes first through a window negative (or a scribing) of the single line, then spreads through a spacer to pass each side of an opaque positive image of the line before reaching the film.

5.8 Process Colour Production

As stated in Section 5.6 all colours can be produced using only four printing inks: the three subtractive primaries and black. Successful application of this technique results in economies in plate making and litho printing. How best to apply it to map production?

Given a printed map it can be photographed through colour filters and four printing plates prepared. However, the material for a new map consists of a set of drawings, all black, usually one for each hue to be produced.

Printers of coloured printings or photographs must generally reproduce the original hues as faithfully as possible; but the map producer has considerable latitude in choice of colours for his final product. This makes the task easier; in particular it is possible to select hues each made up from only two primaries. This greatly eases registration problems. Cyan and yellow in various proportions give a range of greens; magenta and yellow give reds, oranges, and buffs. Mixtures of cyan and magenta can be avoided; so can any shades of colour with black.

For the success of the process, colour quality control throughout production is vital. A hue chart must first be prepared using the same paper, inks, and press as the final map. Densitometers are needed for checking screen percentages. Printed colours must be compared with control strips at each stage. Daylight lamps should be used wherever inspections are made.

5.9 Proofing

In the preparation of a multicolour map, the drawings for each colour plate are made in black ink or are scribed. Although many meticulous checks are made on each plate and on matching of plates (see Section 4.11) before the plates leave the cartographic office, it would be unwise to proceed to final printing of the map without first checking a sample of the product. This is called a *colour proof* and the process of making it is *proofing* or *proving*.

Before the advent of plastic sheets proving was done by transferring the images from the drawings to metal printing plates and printing a few proof copies on paper, using the specified printing inks and a proving press (see Section 5.12.1). The image on the metal plates could not easily be altered, so any required corrections were usually made on the original drawings and new printing plates were then made; altogether a lengthy and cumbersome procedure.

The availability of white opaque polyvinyl and polyester sheets (such as Astrafoil and Stabilene) resulted in a new technique, still widely used, for producing a *dyeline proof*. The first step is to make a positive wrong-reading copy of each colour flap on coated transparent polyvinyl. This is called a *reversal*. It has to be checked for dirt, weak detail, etc.

An emulsion of ammonium or potassium dichromate in a colloid such as gum, gelatine, albumen, or fishglue, is freshly prepared (it does not keep in storage). This is a slow-acting photochemical which under the action of ultra-violet rays causes the colloid to harden and become insoluble. The emulsion is evenly applied (preferably in a whirler) to the white opaque polyvinyl sheet and dried.

The wrong-reading reversal is placed face down over the proof sheet in a vacuum frame and exposed under an ultraviolet lamp. The non-detail areas are hardened but under the opaque detail the emulsion is unchanged and can be dissolved away, leaving bare the plastic sheet surface. A plastic solvent containing a coloured dye is applied to the sheet; it adheres to the bare detail. The hardened coating on the non-detail areas is then removed with citric acid or other suitable solvent.

The first reversal to be transferred in this way is that for the black plate; the dye used is black, and the end-product is a black right-reading image on the white proof sheet.

The proof is recoated with dichromate and the whole process repeated with the reversal of the red plate, this time using a red dye, and so on for the other colours, care of course being taken that each colour plate is in exact register with the black.

The completed dyeline proof can then be sent back to the cartographic office for checking, particularly for bad registration between plates and for detail of one colour clashing with that of another colour.

It should be noted that a dyeline proof cannot be used as a check on the colours of the final printed map, since the dyes used will not exactly match the colours of the specified printing inks. Ink colours should be tested separately — see Section 5.14.

New variations of the dyeline process appear at intervals. For example, a negative reversal may be used instead of a positive. In this case the required dye must be mixed with the dichromate coating. Exposure hardens the coating under the detail and the unhardened coating over the unexposed non-image area is washed away. Although quicker than the positive process it may require more expert processing.

A cheaper alternative to polyvinyl is a stable paper made from man-made fibre. The successive coatings can be wiped on instead of being poured on in a whirler. Kwikproof is the name of a widely-used wipe-on process for working from negatives.

Instead of recoating and drying the proof sheet before exposing each reversal it is possible to use sheets of presensitized transparent foil. Cromalin is the trade name of probably the best-known example of this process. A sensitized sheet of photopolymer on polypropylene foil is laminated in a special machine on to a base sheet, exposed to the positive reversal in a vacuum printer, and the unhardened, unexposed detail dyed with a toning powder in a toning console. The process is repeated for each colour, all on the same base sheet. This is a quick process and the equipment is easy to operate. A later development of the equipment permits the use of negative reversals. The 3M process also produces colour foils from negatives without using a darkroom.

5.10 Development of Printing

Printing was known in China 1500 years ago: the technique was to draw an image on a wood block, carve away the non-image areas, ink the elevated image and transfer this to the material on which the copy was required. The technique became known in Europe but there was little progress until the invention 500 years ago of moveable type, each letter or figure being cut or cast on a separate piece of wood or metal. These could be arranged in a holder to form words, lines, or pages and then rearranged to print another page. The letters are made wrong-reading and are raised (hence the name *relief* process) as in a typewriter, ink being applied to the raised surface of the letters. The paper was pressed down on to the inked surface, hence the term letterpress.

The reverse approach to the problem produced the *gravure* or *intaglio* process in which the image is carved into a block of wood (*woodcuts*) or engraved or etched into a metal plate (usually copper, because it is soft). Ink is spread over the plate and wiped off the non-image areas but remains in the recessed detail from which a print can then be taken. If the plate surface is slightly greasy and the ink is water-based, the ink will not spread away from the detail. For good transfer of the ink to the paper it is usual to damp the paper. The earliest printed maps were made by this method, which was still in use until comparatively recently for printing nautical charts. The image had to be engraved wrong-reading. 'Copper-plate' is a style of writing developed for its suitability for engraving. An alternative to mechanical engraving is to etch the detail image into the metal with acid.

5.11 Lithography

Neither gravure nor relief is suitable for rapid map production; instead of having to raise the image above, or cut it down into, the printing surface, it would obviously be simpler to have it on the surface, i.e. a *planographic* image. The clue to a successful technique was discovered by Alois Sennefelder in Bavaria in 1789 when he found that he could take prints from an image drawn in oily ink on a smooth block of damp porous limestone. This gave the name *lithography* to the process although stone was later superseded by a wettable metal surface.

The stones were of course flat and the image was drawn on them wrong-reading. The invention of photography made it possible to transfer a drawing to a printing plate by a method similar to that already described for proofing (Section 5.9). Finally, metal plates, which replaced stones, are flexible and can be wrapped round a cylinder; this permitted the introduction of modern highspeed printing.

5.11.1 Graining

The basic principle on which lithography is founded is that oil and water do not mix. The process would operate successfully if the plate had a greasy surface, using water-based ink on non-greasy image detail. But in practice it is easier and more satisfactory to use a wet or damp plate surface and greasy ink on greasy imagery.

The first requirement is therefore a plate with a wettable surface. Water will not cling to smooth metal but it was found that the surface of zinc could be slightly roughened so that it could be damped. The roughening process is called graining (or etching). The plate is placed in a tray and covered with glass or steel balls, emery or other abrasive powder, and a little water. A rotating crank attached to the tray shakes it around in a horizontal plane and the resultant abrasive action grains the plate surface.

Plates which have been used for printing one map can be prepared for use on another job by regraining, which removes the old image from the plate surface. However, zinc plates and graining machines have been largely superseded by anodized aluminium plates.

5.11.2 Anodizing

Aluminium plates are thinner (1.5 mm), much lighter, and easier to handle than zinc plates. Instead of graining they are made wettable by anodizing. The plate is placed in an electrolytic cell where it is wired as the anode. The electrochemical action produces a coating of aluminium oxide on the surface of the plate. This has a porous structure which will hold water. The oxide is harder than aluminium but even so the plates cannot be used to print such long 'runs' (number of copies) as zinc plates. However, they occupy less storage space and it is possible to revise the image so they may be retained until the map becomes obsolete. Map production organizations are unlikely to be equipped with anodizing plant so the plates cannot be used again for other maps.

Other plates

For very long runs (say 500 000 copies) bimetallic plates have been designed. They are expensive. They may comprise a copper surface (which will accept oil) on a chromium, aluminium, or stainless steel base (which will accept water). The copper is dissolved away by acid in the non-image areas. For printing-down from negatives, chromium on copper on stainless steel (or zinc or aluminium) is used.

Small plates for short runs can be made from special wet-strength or plastic-based paper. Drawings can be made directly on such plates and letterpress can be added by a typewriter using a special ribbon.

5.11.3 Plate coating

A printing plate is prepared to accept an image, first by etching (cleaning) the surface (dilute acetic acid is used on zinc plates) and then coating with a photochemical emulsion. This is usually applied by pouring it on to the centre of the plate while the latter is rotated rapidly in a whirler, whereby the emulsion is spread evenly over the plate. The coating is then dried.

The active chemical in the emulsion has traditionally been a dichromate, commonly ammonium dichromate, often with the addition of ammonium hydroxide. This is slow-acting, sensitive to strong ultraviolet or violet light but relatively insensitive to visible light so that it can be processed without haste and excessive precaution. The choice of emulsion base fluid depends on the method of plate making to be followed.

An alternative to dichromate is a diazo compound. This has the advantage that it can be stored (whereas dichromate deteriorates and therefore has to be freshly prepared before use), hence plates can be coated at a factory and delivered ready for printing-down. Such presensitized plates (usually anodized aluminium) are generally superseding coating by the map producer.

Diazo cannot be applied direct to bare metal, so the plates used are either anodized aluminium or silica-coated. Diazo in a gelatine emulsion (also some plastic emulsions) hardens the emulsion on exposure to light.

The process of transferring the drawing to the printing plate is called 'printing-down to metal'.

5.11.4 Helio (albumen) process

This was for many years the standard process for printing-down to photolithographic plates. The reversal used must be a transparent negative, and the image produced on the plate is planographic.

The photochemical is mixed with albumen to form an adhesive emulsion. Ammonium hydroxide is included as a preservative for albumen. The dried coated plate is placed in a vacuum printing-down frame under the wrong-reading negative reversal and exposed to ultraviolet.

Light through the transparent detail on the negative hardens the emulsion. The plate is removed and oily development ink is applied; this adheres to the hardened albumen. The still-soluble coating on the non-image areas is washed off and the bare metal etched with acid to increase its water attraction. It is then gummed to desensitize any remaining traces of emulsion and to preserve the bare metal from oxidation. The gum is wettable but repels oily ink. The plate is now ready for the printing press.

The oily-ink right-reading image is on a hardened albumen base raised very slightly above the surface of the metal plate. The image gradually wears away during the printing operation so that there is a limit of, perhaps, 30 000 'pulls' from a helio plate. (By comparison, with a bimetallic plate, the image is oily ink on a copper base on the surface of a steel or chromium plate. The copper base has a working life many times longer than an albumen one.)

Over-exposure during printing-down will produce coarse imagery on the plate but precise timing is not so vital as in some other processes and therefore the operator can be less skilled.

Another advantage of the helio process is that if there are several tints of the same colour, drafted on separate flaps, the negatives can be exposed successively to the printing plate before it is developed, thus producing a single plate for the colour. (However, this method is inferior to combining the flaps into a single reversal which can be inspected and corrected, if necessary, before printing down.)

Instead of albumen, a plate may be given a polymer coating. On exposure to a negative the polymer becomes insoluble in the image areas; it also becomes visible. Further negatives can be exposed until the plate is complete.

Diazo plate coatings have a resinous base. For working from negative originals the resin contains formaldehyde; the exposed resin under detail becomes insoluble in water and is oleophilic while the unexposed coating in non-detail areas can be washed off. For working from a positive original, formaldehyde is replaced by naphthalene; the unexposed coating is insoluble a and the exposed non-detail can be removed.

5.11.5 Deep-etch (gum-reversal) process

All the above methods of printing-down produce a planographic image on the surface of the plate where it is vulnerable to gradual erosion during the ensuing printing run. This can be avoided if the image is produced below the surface of the plate. One way of achieving this is to use the 'deep-etch' process, also known as gum reversal. Despite the name, the etching only goes 0.01 mm below the surface; however this is deeper than normal etching which is simply a cleaning or roughening of the plate surface.

Dichromate is mixed into an emulsion with gum (usually bitumen and chlorocresol). The dried plate is exposed to a positive reversal, and the non-image areas are hardened. The plate is wiped with a violet dye and developed by dissolving away the unexposed gum. The image now shows as bare metal against the violet background. The image is etched into the plate surface by applying an acid solution (which does not penetrate the hardened gum background). Careful timing is important to get the etch the right depth and width. Etching is stopped by washing with alcohol and shellac, then the plate is dried. If the plate is zinc it is now coated with lacquer which enters the etched image; oily ink is added and this adheres to the lacquer.

The ink is cleaned off the hard gum coating which is in turn removed with citric acid. As in the helio process, the bare metal is etched with acid and gummed. The right-reading positive image is now on oily ink on lacquer based below the plate surface so that it will not wear away during long runs. The length of run is limited by the graining on the non-image areas wearing smooth so that it is no longer uniformly wettable.

If a bimetallic plate is used instead of zinc the top wettable non-image surface will be chromium, with the image etched down to oil-receptive copper.

5.12 Lithographic Printing Presses

In the early days of printing, ink was applied by hand to the plate and the paper was pressed down on to the plate by a capstan screw, one sheet at a time. The plate image had to be wrong-reading. When rigid wooden blocks or stones were superseded by flexible metal plates, the latter could be wrapped round a cylinder and rolled across the paper but it was difficult to ensure an even contact between hard metal and paper. This problem was solved by the introduction of an *offset* cylinder: the image is rolled off the metal plate on to the surface of a rubber 'blanket' fixed round a cylinder which in turn transfers the image by rolling over the paper. The compressible rubber blanket makes a good contact with both plate and paper without use of excessive pressure; this also results in reducing wear and extends the useful run of the plate. Since the image is twice reversed by contact transfers, the image on the plate must be right-reading.

5.12.1 Proving presses

Proving presses (fig. 5.4) generally operate with a flat plate and a rotary blanket. They are simpler and cheaper that fully rotary presses and can be used not only for making paper proofs but also for short runs (limited editions) of a finished map Although slower in operation than a rotary press, enough time is saved at the 'make-ready' stage for a flat-bed to be quicker in total time for a run of a few hundred impressions.

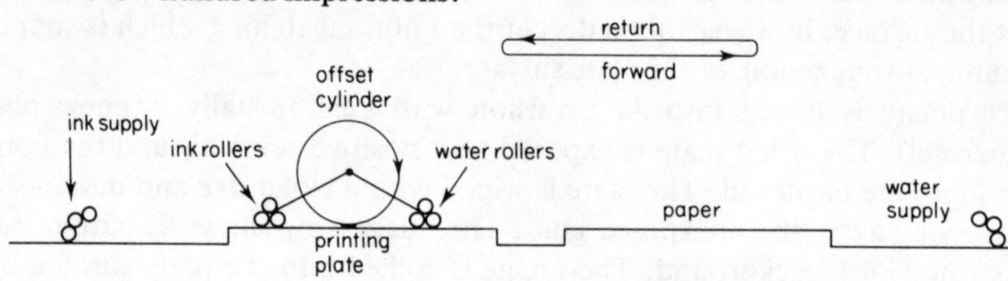

Fig. 5.4 Offset proving press

The plate lies flat, face up, and is automatically damped and inked by small rollers. The offset cylinder rolls across the plate, the rubber blanket picking up the inked image, and then rolls across the paper, printing the image. The cylinder then rises automatically to make the return journey out of contact. The paper feed may be manual or automatic. After printing one colour the blanket and ink rollers must be washed and the plate and ink changed ready to print the next colour.

5.12.2 Rotary presses

In a single-colour rotary press (fig. 5.5) the plate is wrapped round the plate cylinder. As the cylinder rotates, the plate passes under damping and inking rollers, then transfers the image to the offset roller. Paper sheets are pressed against the blanket by an impression roller, the image thus being transferred to the paper. Output is at the rate of perhaps 6000 sheets an hour.

A series of such single-colour presses can be combined to form a mutli-colour press; the paper is fed from one impression to the next and any number

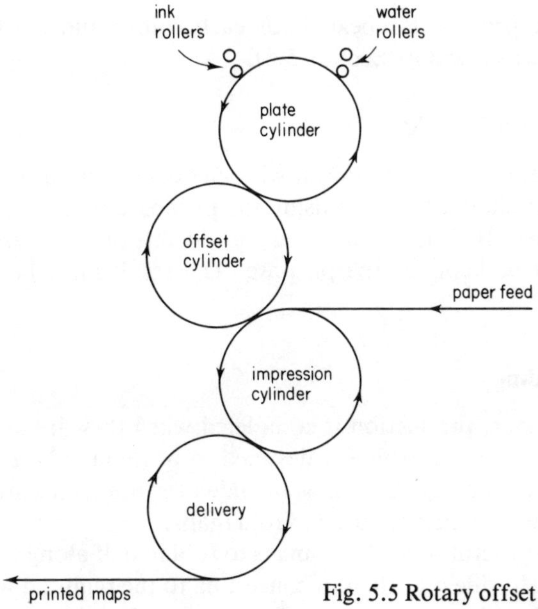

Fig. 5.5 Rotary offset press

of colours can be successively printed. In practice, two colour and four colour are the most common; obviously the latter is well suited to 'process-colour' printing (see Section 5.8).

5.13 Silk-screen Printing

This is a technique which can be used to produce short runs of coloured maps less expensively than by lithography. The principle is to get a negative on a screen to form a stencil. The non-image areas are impervious to ink but in the image areas ink can be forced through the screen on to paper. The screens were originally of woven silk, later superseded by fine wire mesh; today they are usually of woven polyester with 120–140 threads per centimetre.

Various methods of converting the drawing into a stencil on the screen are possible. Currently it is usual to start by transferring (in a vacuum frame) a positive transparent original on to screen-printing contact film, in which the emulsion is sandwiched between a 0.5 mm thick transparent polyester base and a soluble protective coating. The image passes through the base and after exposure the coating is washed away leaving a wrong-reading negative image on the base. The image is pressed on to the diazo-sensitized screen and the polyester base is peeled off, leaving a right-reading negative image (stencil) on the screen. The non-image areas can be damped to repel ink.

The screen is inked and laid over a sheet of paper held on a vacuum paper bed. A hand- or machine-operated squeegee forces ink through the image stencil on to the paper. More ink is transferred than in lithographic printing, so that sheets cannot be stacked until they are dry (some machines have a built-in drier). The process is repeated for each colour and the image quality is similar to process printing (Section 5.8).

5.14 Testing Printing Inks

Before starting a production run with any equipment using printing inks, a test for colour should be made using the proposed inks on production paper (or other material). It is not necessary to do this on the press; a 'printability tester' should be kept for the purpose. The result must be examined under white light.

5.15 Map Folding

For some maps, production is completed when they are delivered from the printing press; they are probably destined to be mounted on walls or laid flat or hung in a map library. But maps intended for use on the ground are usually folded. Special machines are used to fold maps.

The simplest useful style of folding is to fold in half along the longest axis of the map, printed side out, then to concertina to the required width. If the map is large a final fold may be needed. This format allows any part of the map to be seen with a minimum of opening. It also offers scope for ingenuity in the attachment of stiff covers.

An alternative to folding in half, face out, is to print the whole map on both sides of the paper, with norths at opposite edges of the paper, which is then cut in half so that each half includes a complete map (maps destined for this treatment are best drawn in two halves with a blank strip along the line of the intended cut). It may be mentioned here that printing machines are obtainable which will print both sides of a sheet simultaneously (a technique known in the trade as 'perfecting').

Stiff covers for a folded map have to be printed separately and stuck on to the maps by a special machine.

Instead of having a separate cover, a title panel may be printed on the back of a map, which is then folded face inward in such a way that the panel appears on one of the outside folds.

5.16 Paper

'Paper is by nature a perishable material' (*Libr. Qrtrly*, Jan. 1970).

Nearly all maps are printed on paper but in fact this is far from an ideal material for the purpose, its main merit being comparatively low cost. It will disintegrate when wet, fall apart along folds and is easily torn by careless handling. Maps can be successfully printed on strong waterproof opaque plastic sheets, but these cost much more than paper and show little progress in competing with it.

Paper is also used for air-photo prints, for drafting and tracing in the drawing office, and in most copying systems. The compleat map producer should therefore have some knowledge of paper quality and its effect on his work.

Paper is basically formed from natural cellulose and its strength depends on whether the latter is long fibre or short fibre (man-made fibres are now used in making some papers). Soft woods (e.g. spruce and pine) are the main source of short fibre while long fibres come from esparto, sisal, and rags (which themselves are originally made from cotton, flax, etc.).

The raw material is pounded and wetted to make a pulp and then treated with chemicals to soften and separate the fibres (and to bleach them if the product is to be white). The fibres are translucent and when the pulp is dried they form a matrix with air spaces. This structure gives paper compressibility and resilience, important for printing and folding. The spaces absorb the fluid bases of inks, leaving the pigments on the surface. The pulp has to be sized with rosin to reduce this absorption, otherwise the product will be blotting paper.

To make tracing paper the already translucent pulp is further beaten to a gel which is transparent. Tracing paper is needed in thin sheets so it must be made of long fibres to retain enough strength.

To increase opacity, white fillers such as talc and kaolin are added. For card, starch is added as a stiffener. The treated pulp is calendered, i.e. pressed, dried, and rolled out in a continuous sheet of the desired thickness. The product is initially in large rolls which can later be cut into sheets of required sizes. A ream of 500 sheets is the standard measure of paper quantity.

Thickness is measured by weight: litho-printing papers are usually in the 80–200 grams per square metre range. In *web-offset* printing machines the paper feed is direct from rolls instead of sheets and it is printed on both sides.

Short-fibre papers are the cheapest, e.g. newsprint and cartridge. Addition of long fibres improves strength, giving drawing paper (which is extra thick to ensure opacity) and 'rag-litho' used for lithographic printing. Since oil-based inks do not soak into paper so much as water-based inks, less or no sizing is required.

Paper used for photographic prints and other wet copying processes must not disintegrate in water. Formaldehyde is added to the pulp to give it 'wet strength'.

Paper made from man-made fibres has many of the qualities of plastic sheet. However it may be cheaper to use and wear out two or three copies of a map printed on standard paper than to use one copy printed on a more enduring material, such as a plastic or plastic-coated paper.

5.17 Other Copying Methods

5.17.1 Photostat

The process camera (see Section 5.3) produces a copy in two stages: first a wrong-reading negative on transparent film, then a right-reading positive. For some purposes a right-reading negative on paper is sufficient. The photostat and similar equipment will produce this with a single exposure and processing.

The opaque original (maximum size 75 × 100cm) is placed on an illuminated horizontal copyboard. The light rays from this are reflected by a prism or mirror set at 45° in front of the camera lens (which has its axis in the normal horizontal position). Thus there is double reversal and the image on the camera screen is right-reading. The emulsion is on a roll of paper and after exposure, the exposed length is cut from the roll and processed without removal from the equipment. The end-product is a right-reading monochrome negative on paper with a maximum size of 45 × 60 cm. There may be some distortion of the paper, this being a wet process. Adjustment of the lens permits limited reduction or enlargement.

The negative print can be placed on the copyboard and the process repeated to give as many positive copies as required. This is a comparatively cheap but slow process for making monochrome copies at the same or different size of small maps or parts of them.

5.17.2 Microfilm: see Section 8.6

5.17.3 Diazo contact prints

The earliest diazo prints were called sunprints because the sun was used as the source of light (or ultraviolet). They are now usually known as dyeline prints because the addition of dyes to the photochemical gives a choice of image colour: black, sepia, blue, red, etc.

The original to be copied must be a positive image on a thin transparent material (formerly tracing paper or tracing linen, now usually plastic). Exposures may be made in a printing-down vacuum frame but in more modern equipment the original and the diazo paper pass in contact through pairs of rollers and are exposed to ultraviolet from a tube source. The maximum paper width is about 125 cm but there is no limit to the length of the print. Ultraviolet decomposes the diazo salts leaving bare paper in non-image areas. Development of the unexposed image is by ammonia, usually applied by passing the print through rollers damped with ammonia. Final fixing is by hypo or a stabilizer powder to remove any remaining active diazo. The copy is a positive. Instead of paper the diazo may be based on plastic but of course this is a more expensive product.

To obtain a right-reading copy the original is laid on the diazo paper face up and exposure is through the thickness of the original material. This is all too often not perfectly clean and transparent with the result that the non-image areas on the copy are rarely pure white but show a varying density of shades. Copies gradually fade when exposed to light.

5.17.4 Blueprints

Ferro-prussiate (basically potassium ferricyanide) has long been used as a photochemical. Exposed areas turn blue. Unexposed salts are washed away leaving a negative white image on a blue background, commonly called a blueprint. It is the easiest way of getting a positive print from a transparent negative original.

5.17.5 Xerography and Electrostatic copying

For small copies these are now the most widely used processes. Xerography is an entirely dry process and hence much quicker than any other copying system. There are several variations of the basic process, in most of which the image passes through a lens so that reduction and enlargement are possible. For document-copying work the lens is usually fixed and produces same-size copies but for map work enlargement from microfilm (see Section 8.6) is a standard procedure. The greater the desired enlargement, the more expensive the lens must be. It is rare to find final copies larger than 50 × 75 cm.

The (opaque) original is laid face down on a glass plate and kept flat by a heavy rubber cover. It is briefly illuminated from below, usually by a moving (scanning) strip of light and the image passes through the lens to fall on the surface of a rotating cylinder (or continuous belt) which is coated with selenium or zinc oxide. The coated surface has passed close to an electric wire and acquired a static charge of 6000 volts (positive).

The action of light in the non-image areas discharges the static charge; the dark (wrong-reading) image detail retains the charge, which then picks up negatively charged black powder. As the cylinder or band rotates it comes in contact with the positively charged coating of a sheet of white paper to which the negative powder is attracted to form a right-reading image. The surface is heated to fuse it to the powder. Each rotation of the cylinder or band produces another copy.

Obviously the charge sequence described above could be reversed, i.e. negative cylinder, positive powder, negative paper.

If the original is a diapositive or film positive it enters the printer face up and lighted from above. Exposure is through the lens directly to charged paper on which an enlarged right-reading image is formed.

The direct electrostatic process omits the intermediate cylinder and includes a mirror or prism in the optical system to give a right-reading image directly on the output copy paper. Instead of charged powder fixed by heat, a liquid toner is used and fixed by drying with an airblower. The system can be adapted to rapid printing direct from computer tapes (see Section 7.16).

It is also possible to adapt these processes to preparation of printing plates, by using a suitable plate as base for the output copy, water-receptive electrostatic plate coating, and ink-receptive image-forming material.

New copying techniques (or variations of existing techniques) frequently appear, each representing another step towards producing the perfect system which would have the following attributes. It would be simple to operate (not requiring a darkroom); distortion-free (therefore a dry process); able to produce single or multiple copies rapidly and to reproduce multicolours, reductions, and enlargements; work from opaque or transparent positive or negative originals; and produce at choice copies that are opaque or transparent, positive or negative. And of course it must be cheap! Clearly these requirements leave plenty of scope for research and development.

6

Map Revision

6.1 The Changing Face

Topographical surveyors are sometimes asked 'What will you do when you have finished mapping this country?' The answer is of course that almost every map is out of date before it is published. Not only do man-made developments such as new buildings, roads, and other works continually appear, but even in uninhabited areas natural processes raise and lower the land surface, cause landslides, change river courses, spread deserts, and so on. If current maps are to give a reasonably up-to-date image of the environment then revised editions are frequently and repeatedly needed.

6.2 Print-run

When any map is sent for printing publication copies, the number required must be specified. Several estimates have to be made: What will be the rate of demand? How long will the stock last? When will a new edition be needed? What is the rate of change of surface detail in the area? If the map is not a first edition, then inspection of earlier editions, their publication dates, print-runs, changes in detail between editions, provide enough clues to make a reasonable assessment. For the first edition of a new basic map of an area, the history of a sheet of a similar area may suggest the right answer.

6.3 Reprint, Revise, or Reconstruct?

When the remaining stock of a printed topographical map is down to less than the average demand over one year it is time to consider reprint or partial revision.

Sometimes there is a requirement for a completely reconstructed map, for example if the map is uncontoured and material is now available for a contoured edition, or if contours are in feet and a metric edition is required, or if detail has become so dense that a larger-scale map is needed. Any of these might take several years to organize and produce; meanwhile a reprint of the existing edition will be the simplest stop-gap. If there has been little development and change in the map area, a reprint will probably be adequate.

6.4 Updating of Charts

For navigational charts, particularly nautical charts, it can be of crucial importance that they are up-to-date: a new uncharted danger might cause the loss of a ship valued at many millions. Hence procedures are not the same as for maps.

Urgent corrections can be broadcast by local radio; they can be repeated in weekly printed notices sent to chart-holders whose names are on a mailing list. If the corrections are too complex to describe in words, the chart publisher may rush production of a revised copy of a small area of a chart which can be stuck on to existing editions.

Unissued stock of charts must be brought up-to-date by hand (in respect of such amendments) before issue; hence print-runs tend to be short and stocks are kept small.

6.5 Revision Sources

There are many sources of revision material. For maps the most effective is a new set of air-photos of the area, exposed and printed at map scale. The cartographic office should possess a projector (there are many designs available) in which an air-photo is placed and its image is seen (actually or apparently) superimposed on a copy of the existing map. The photo can be tilted to compensate for tilt of the camera at the time of exposure and quick adjustment of object and image distances can make the local scale of the photo image the same as map scale. Then visible changes in detail can be drawn directly on the map or on an overlay. Unchanged detail is superimposed to control positioning.

Another revision source is engineering and other plans showing the layout of new roads, works, housing estates, etc. These usually require reduction of scale and keying to identifiable old detail.

A field check of the revised work is highly desirable because there are always changes which do not show on air-photos and new construction is not always exactly or completely executed in accordance with design plans.

Most topographical maps carry a printed invitation to users to report errors, but few users respond. Nevertheless, a revision record copy should be kept for every sheet, on which are entered notes about any amendments which become known during the currency of the existing edition.

6.6 Revision Drawing

When most maps and charts were printed from engraved copper plates, revision was made directly on the plates. But metal plates are bulky to store and may be required for regraining and printing other maps. Diazo-coated anodized aluminium plates can be revised by deleting with phosphoric acid and adding new detail with cellulose lacquer. Deep-etch plates cannot be revised.

The general procedure is to revise images at a point further back in the production line and to make new plates.

For some of the colours in a topographical map little or no revision may be expected. For example, relief rarely changes measurably, so that the flaps for contours, layer tints, hill-shading, etc. need no action. Blue water plates may have a few new pipelines and reservoirs, green vegetation plates may have new plantations. Most of the change is man-made and is to be shown on the black and red plates.

Deletions should be effected first and this is most easily done by opaquing on a negative.

The essential requirement for *additions* is that they will be in register with the old work. There are several possible methods. Generally it is best to plot a compilation and print it in blue or other colour in register as a key on a blue (preferably made after deletions: see above) of old work. The latter may be in positive form for ink drawing or on scribecoat for scribing the additions. The fair drawing (or scribing) of the new work can later be combined with the old work to produce a revised reversal from which the printing plate can be made.

It is possible to scribe new line detail on a yellow scribecoat, add names and symbols in black, and transfer both to autopositive film by successive exposure to yellow and white light; on development a combined positive image is obtained.

7

Computers and Cartography

'Human error is nothing to what a computer can do if it tries'
(AGATHA CHRISTIE (1974) *Halloween Party*, Pocket Books, N.Y.)

7.1 This Computer Age

Computers have already influenced and will continue rapidly to invade every technological activity, including surveying and map production. As preceding chapters have shown, map production is a series of processes some of which already obviously benefit from computer assistance whereas for other processes the applications are still at the experimental stage.

Whenever any new equipment, material, or technique is contemplated, designed, or announced, the testing questions are: Is it quicker? Is it cheaper? Does it give a better result? The answers given in different parts of the world may not be the same. Where capital is scarce, labour cheap, and unemployment high, computers may not be very welcome and maps will continue for a few, or many, years to be produced by the traditional methods already described.

Whatever attitude is taken towards computer aids in cartography, one thing is certain and that is that they cannot be ignored. Therefore present applications in map production and probable future developments must now be considered.

Electronic computers manipulate numbers in digital form; this requires that all data and all commands to be fed into the computer must finally be in numerical form (whereupon the first thing the computer does is to translate decimal-base figures into binary-base numbers). Large computers have 'compilers' and other peripheral 'software' which can translate an alphabetical computer 'language' (Fortran, Basic, Algol, Pascal, etc.) into digital form so that commands and program may be entered through a typewriter keyboard. The operator has first to learn the 'language' to which the computer is receptive.

Smaller computers have 'hard-wired' permanent programs where pressing one or two keys will put the program into operation, e.g. convert rectangular coordinates to polars.

Large computers can obviously perform more complex tasks than small ones, but to get the maximum return from their higher cost they are often connected by wire to a large number of user 'terminals' and operate on a 'time-sharing' basis. This may be satisfactory for occasional mapping operations such as computing the grid values of graticule intersections but not for larger specialized tasks such as drawing a map. Satisfactory map production requires 'dedicated' computers, specially designed for the tasks to be performed and used solely for those tasks.

7.2 Map Data Capture

Computer operation can be broadly divided into three main stages: data capture and storage, data processing, and data retrieval.

In map production, data capture implies converting topographical and other map detail into digital form, and is commonly known as digitizing.

Any point on a map which has known coordinates is already in digital form; grid intersections, survey control points, etc. The coordinates may be on different systems, e.g. graticule intersections in geographicals (latitude and longitude) and other points in projection grid values (probably metres), but if the mathematical relationships between the systems are known the computer is easily programmed to convert all to the same system. Hence there is already enough digital data to enable a computer-controlled plotter to construct the framework of a map.

For new basic mapping the best time to collect digital data from map detail is at the photogrammetric plotting stage. The equipment has (x,y) scales and drives a coordinatograph which draws the map-plot. All that is needed is a means of continuous recording of the (x,y) values of the position of the floating mark or plotter point on the table. Such data can be recorded on magnetic tape or disc or paper tape. It is also necessary for the operator to be able to enter a numerical code to identify the feature being plotted.

Heights are best recorded separately, either as third coordinates in a list of heighted control points or as a constant (z) value for a string of (x,y) values defining the map position of a particular contour.

More frequently, the input for a topographical data bank will be digitized from existing basic maps. Obviously it should be compiled from the largest available scales which show the most detail. If the map is to be manually digitized and is smaller than the digitizer table it is advantageous to enlarge it to table size. The effect of any small errors in digitizing is proportionally reduced. In practice it is easier to digitize negatives; and for colour maps to digitize a separate flap of each colour.

7.3 Digitizing Precision

Digitizing equipment is designed for converting the graphic image on the map into a string of coordinate values. Map detail may generally be classed (see Section 2.2) as point, line, or area. Digitizing a point is no problem: an easting and a northing (or an x,y value) will fix the map position of any point. The precision required in measuring and recording these coordinates depends on the scale at which the data will be replotted. In normal practice this would be at the same or a smaller scale. The limit of plotting precision is generally accepted to be about 0.1–0.2 mm hence there is little to be gained from trying to record position more precisely.

Each digitizer has its own datum or zero point. For equipment which operates on the principle of measuring rectangular Cartesian coordinates this zero point will normally be near the lower left-hand corner of the digitizing table and just outside the southwest corner of the map to be digitized. If the table is one metre square, then to fix a point to 0.1 mm precision, (x,y) or (E,N) must be measured to four significant figures in each direction. The fourth figure is likely to vary somewhat if the measurement is repeated.

A straight line is simply defined by two points. A regular curve can be reconstructed from three or four points. But the vast majority of lines on maps are made up of irregular curves, which can only be digitized as a string of points. How close should the points be? It would be possible to record a position at, say, every 0.2 mm along a line but in the straighter sections of the line a high proportion of the points would be superfluous. For economy in time and labour only the essential minimum of fixings should be made. As a rule-of-thumb, the spacing between points should be proportional to the local radius of curvature of the line, that is, far apart on sections which are nearly straight, but close together in sharply curving sections.

Suppose that on a sharply curving line an eight-figure coordinate is recorded every millimetre, then to define a centimetre needs 80 decimal digits (and many more binary digits). A contoured map may have hundreds of metres of lines; the number of 'bits' (binary digits) to be recorded for a single map sheet can clearly be very large. However, magnetic tapes and discs have a large capacity (measured in megabytes, i.e. millions of groups of bits) for storing such data.

An area on the map is defined by digitizing the perimeter line, starting and finishing at the same point. The starting and finishing values may be compared as a check on the performance of the digitizer.

In addition to the string of digits defining position on a map of an item of detail, it is also necessary to enter a numerical code identifying the detail, e.g. school building, 200 m contour, forest boundary.

7.4 Digitizing Equipment

Digitizers may be designed to measure (x,y) coordinates or polar coordinates, the datum in each case being the digitizer zero point.

It might be thought that an automatic coordinatograph (see Section 4.5) could be adapted to work in reverse: the plotting point being placed over the point to be digitized, the (x,y) value is displayed and recorded. There are some digitizers of this type (usually with servo-motors to aid movement) but it is not the easiest equipment to use for line following.

One variation places the moving mechanical equipment under the surface of the table; it is caused to move by receiving electric impulses from a coil or electronic pencil tracing over the map detail on the surface of the table, (x,y) being measured and recorded as before.

Any sort of mechanism must have weighty moving parts with intrinsic inertia and slow reaction. Hence this type of digitizer is likely to be superseded by a design with no moving parts. A very close (x,y) grid (with 0.1 mm or closer spacing) of conductor lines known as a *datagrid* can be obtained by modern micro-printing methods and set in the surface of the table, and automatically records the (x,y) position of the centre of the digitizing cursor.

Another principle used is that of the transducer, or magnetostrictive ranging, by which the distance of the cursor from the x and y axes is sensed by electronic radiation. No grid is needed and the equipment can be operated on a metal-surfaced table.

One type of polar digitizer has the cursor at the end of a radial arm. The arm slides through a slot in the rotary measuring head (which is the zero point for coordinates) and the distance of the cursor from the zero point is automatically measured and recorded. At the same time the angle between the arm and a zero direction is recorded. The recorded polar coordinates can be converted by computer to rectangulars. Since the radial arm and the cursor can project in any direction from the zero centre, the greatest area can be covered by placing the rotary head in the centre of the map and not at the SW corner. Some of the polar coordinates will then convert to negative rectangulars. See fig. 7.1.

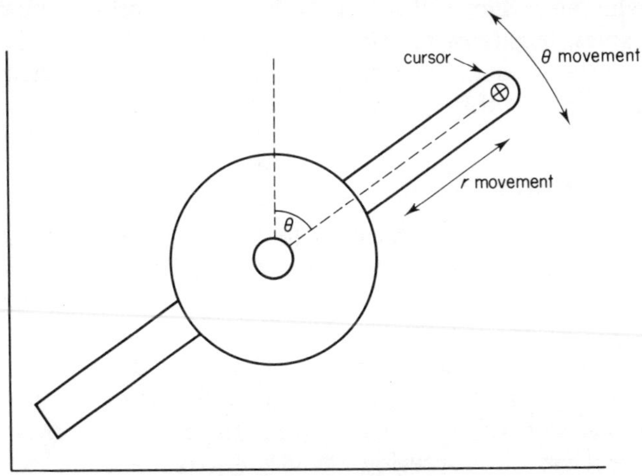

Fig. 7.1 A type of polar digitizer

Another type of polar digitizer (the Aga Geotracer, 1975) does not measure any distances but instead measures two angles. The design resembles a human arm; the shoulder joint is the zero point and the cursor is at the hand. The angles at the shoulder and elbow are automatically recorded. The radial distances are fixed and a computer can convert the measured angles to (x,y) coordinates. Again, for maximum area cover, the zero point should not be in the SW corner of the map. See fig. 7.2.

A typical arm length is 42 cm and angles can be recorded to 1/5000 of a right angle (about 1′). This gives an accuracy better than 0.2 mm. This pattern of

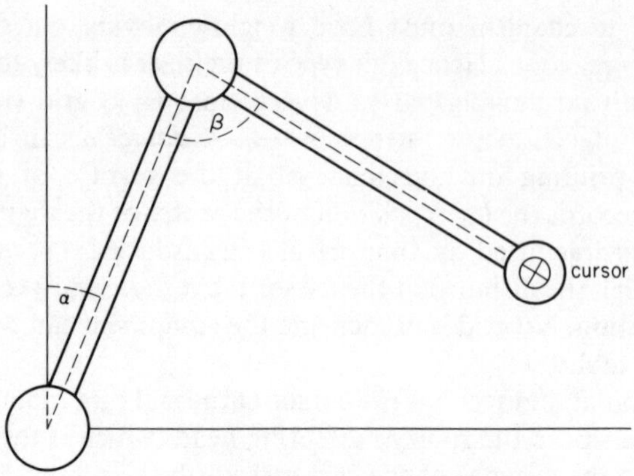

Fig. 7.2 Another type of polar digitizer

digitizer is obviously more compact and portable than one which requires an electronic table.

The cursor of a digitizer is usually a cross-hair under a magnifying lens; this can be positioned accurately over the detail. It may carry a small display tube which continuously shows the (x,y) values of its present position. Some digitizing tables, mostly of the electronic grid type, can be tilted so that the operator can easily reach any part. On such tables the cursor is usually attached to some kind of flexible gantry for support.

7.5 Digitizer Operation

A digitizer is normally connected to a control unit which includes a keyboard and frequently a CRT (cathode ray tube) display, whereby instructions can be given to the digitizer and connected computer and information can be received from it. In computer jargon this is an interactive mode. For instructions that recur frequently there are usually several push-switches on the cursor: one may instruct it to record the coordinates of the point under the cursor, another to suspend recording while the cursor is moved, another to repeat the label code of the previous detail and so on.

7.6 Digitizing Lines

7.6.1 Manual digitizing

As previously mentioned, there may be in one map sheet hundreds of metres of lines, largely made up of irregular curves, the conversion of which to digital form absorbs a large proportion of the total effort required to record the graphic image in numerical form.

In some equipment, the cursor can be set to record its position at constant time intervals (say, every second), or at constant distance intervals (say, every millimetre). Although it is possible to vary the intervals both these modes have disadvantages and in practice most operators prefer to record at positions chosen by themselves. (Note: for features consisting of two or more parallel lines, only one line need be digitized. The computer can be programmed to draw the other(s).)

As in other tasks requiring visual judgement (such as pointing and reading a theodolite) there must inevitably be more or less human error in placing a cursor exactly on a point of map detail, and such error is multiplied in tracing along a string of points which make up a line.

It is evident that time should be saved and error reduced if the human operator is assisted by an automatic line follower.

7.6.2 Automatic line following

Various techniques have been devised and tested; currently the most effective is a modification of the scanning system (see Section 7.14). The flap to be digitized (e.g. a contour plate) must first be reduced to negative form and area not exceeding 96 × 68 mm on an A6 microfiche film. The width of lines on this negative must not be less than 0.03 mm. This may limit either the reduction factor or the area of the original which can be digitized in one operation.

The negative is scanned by a very fine helium–neon red laser beam and the scanning area is automatically restricted to a narrow zone containing the line to be digitized. The beam scans parallel to either the x or the y axis and automatically changes from one to the other as required by the direction of the scanned line, e.g. a west–east line is scanned by a vertical (north–south) scan pattern. At each encounter (crossing the scanned line) the beam passes through the open line on the negative to a photoelectric cell; the line width and position of the line centre is transmitted to the computer.

Encounters occur at 0.010–0.015 mm intervals along the scanned line (a far closer spacing than is possible with manual digitizing) but at the rate of 500 encounters/second digitizing proceeds rapidly.

At the same time the negative is projected with × 10 enlargement on to a display screen (Fresnel screen), size 100 × 70 cm, so that the operator can monitor the scanning. The control console includes a joystick or tracker ball whereby the cursor (a laser-generated cross on the display screen) can be moved to the start-point on the feature to be digitized. The scan will automatically stop at the end of a line or may be stopped by the operator.

Progress check

At the end of a line the digitized data can be 'played-back' and plotted by an argon-ion blue laser beam which draws a black line on photochromic film (sensitive to blue but not to red) overlaying the negative. This 'paints-out' the line on the operator's screen. The image on photochromic film fades after a time (usually about twenty minutes) so that the film can be usually repeatedly. The operator can assist the scanning with guidance at junctions and over gaps and ensure that the scan does not transfer itself to a closely adjoining line. He also has to enter feature codes.

7.7. Labelling

Whenever the position of a feature on a map is recorded, the nature of the feature, and perhaps also its name and size, must also be recorded. This is somewhat similar to compiling a gazetteer (Section 2.5.5) and it may be possible to use the same feature codes for both.

First must be compiled a list or glossary of all types of feature which may be found on maps to be digitized, and each entry given a five figure feature code. It would be possible for each type of feature to be represented on the control unit keyboard by one key, the pressing of which would enter the feature code on the tape. This would be an uneconomical design compared with a standard keyboard on which the operator has to press several keys for one code. However, the latter requires either that the operator frequently consults the glossary or attempts to memorize the codes, of which there may be a hundred or more.

A more efficient system uses a *feature menu*. A part of the digitizing table, normally near the right-hand edge, is reserved for the menu (or this may be on a separate table or mounting). The menu consists of a list of the features possibly represented by pictograms or other symbols for rapid identification, and each in a rectangular frame or box. If the cursor is placed within any frame and the appropriate button is pressed, the control unit will translate the table coordinates of that point into the correct feature code. The process is simplified and speeded up if a second cursor can be used on the menu so that the main cursor need not be removed from the area being digitized.

Still more advanced equipment uses 'voice recognition' by the computer. The operator only has to press a switch and say the name of the feature (e.g. 'lighthouse') or perhaps a height, and the code is (hopefully) recorded.

Instead of recording frequent changes of feature code, there are obvious benefits in recording all items of one type of feature consecutively. For multicolour maps this is partly achieved by digitizing each colour flap separately. It is also necessary to adopt a procedure to ensure that no item of map detail is omitted from digitizing. A practical method is to make a 'pre-edit' print of the flap to be digitized and on it write a serial number against each item in the order each is to be digitized. Such serial numbers facilitate rapid recovery from computer data storage. Some features may need two codes, e.g. a road or river which is also a boundary.

7.8 Checking and Editing

Errors and omissions will inevitably occur in digitizing and it is therefore essential to check the data record. One technique displays the digitized detail on a raster (cathode ray tube) as the work proceeds. Storage tubes will retain the image up to about sixty minutes while in 'refresh' display tubes (e.g. the Tektronix) the input has to be repeated at least thirty times a second to give a flickerless image. The work station (interactive control unit) has a facility for the operator to delete errors and amend the data file. Another method is to run the data tape or disc in a fast automatic plotter to produce a same-scale 'edit plot' which can be compared with the original and used as a basis for amendments. These can be recorded on a correction tape which is then run in the computer with the original data tape to produce a combined revised tape.

When changes in detail on the ground occur at a later date, the revision data can similarly be entered into the data record.

7.9 Geographical Names

The digitizing operations outlined above record (x,y) coordinates plus feature labels on paper tape, magnetic tape, or disc. When this record is replotted it will display all the points and lines, in various symbolic form, which constitute the topographical map detail, but no alphanumerics (letters and figures). The majority of these are geographical names which, for various reasons, are best dealt with separately.

A major problem with names is positioning: the name of a point feature may be printed around the point, while the name of a line or area feature is often spaced along the feature, not necessarily east–west or in a straight line.

It is found most practical to record each name on a separate computer card, on which additional information can be added as desired, e.g. population of a town, its administrative status, etc. The punched cards are then ready material for preparation of a gazetteer (Section 2.5.5).

7.10 Data Bank

On converting any graphical image (air photo, plot, or map) into digital data, the first item recorded is a reference (e.g. series number, sheet number, edition) by which the data can later be identified. Marginal information round a map generally need not be digitized; most of it can be recorded through a keyboard. For series maps, data for one standard margin should serve for all the sheets.

Digitizing equipment measures table coordinates referred to the zero axes of the equipment. The data file needs the coordinates to be on the local map grid system. Therefore digitizing must start by recording a set of points (e.g. the four corners of the map) whose map grid coordinates are known. The computer is programmed to convert all recorded table coordinates into grid coordinates giving a best mean fit over the set (which may be as any as 16 points).

Temporary data storage in a computer is usually on a magnetic disc

(400 × 10^6 bits). This is a 'direct access' store which is expensive but allows rapid retrieval of data. Long-term storage (the data bank, data base, or data file) is on 'floppy disc' (capacity 300 000 characters) or on a diskette or on magnetic tape (sequential storage, 1250 bits per cm). A single map sheet might be recorded on 10 to 50 m of tape. Although tape length may be as much as 2 km, it is convenient to record the whole of one map, and one only, on one tape. A 45 m tape cassette (180 000 bits) is often suitable.

The information stored in a data bank should be updated (revised) as often as possible otherwise it rapidly loses its usefulness and value.

7.11 Data Processing

Data processing is a task for computer technologists rather than for cartographers. However, the map producer should know the capabilities of processing and the results obtainable.

The computer operator can write computer programmes which are transferred on to punched cards, magnetic tapes or magnetic cards, commanding the computer to process the data on the data tape in various ways and to display the results. One mapping task effected quickly and efficiently with computer assistance is the drawing of grids and graticules. Selected data can be quickly recomputed on to a different grid or graticule or scale. The computer can be programmed to pick out from the data tapes and draw only certain areas (known as *windows*) or selected classes of detail (identified by their labels), e.g. airfields, or power lines. This can be very useful in the production of special maps, for which thematic information can be recorded on separate tapes.

Change of scale almost always means change to smaller scale, e.g. data digitized from 1 : 50 000 maps to be drawn at 1 : 200 000. This change requires the introduction of generalization, selection, and exaggeration, which all require more critical judgement than can be programmed into a computer. In the example mentioned above there is a four to one reduction in scale; to generalize line detail the computer can be programmed to select every fourth point on the data tape for plotting. This method would inevitably fail to plot some points on the line where there are important changes of direction. Interactive control is needed here so that the plotter-operator can intervene to alter the computer plot.

7.12 Automatic Plotting

Digitizers convert graphical analogue data into digital data and some kind of plotter (ADM = automatic drawing machine) is required to convert the digital data back to a graphic image. It was noted in Section 7.8 above that we can get an instant image in a CRT, and take a hard copy of it, but in practice the size is limited and the distortion is too great to be acceptable. Most automatic plotters are mechanical, operating like the automatic coordinatograph mentioned in Section 4.5 but it is obvious that an accurate electronic plotter with no inertia and no moving parts should be able to operate much faster.

In some (x,y) mechanical plotters the plotting point only moves in the x direction, while the paper (or other material) moves in the y direction. The paper may be either on a flat carriage or round a rotatable drum. The drum plotter is usually faster but its accuracy tends to be lower, and accuracy is more important than speed in map production. In practice most map plotting is done on a fixed flatbed with the plotting point moving in both x and y directions.

The size of such plotters varies from small 'graphplotters' controlled by a 'table-top' size dedicated computer up to machines as large as two metres square. Although maps and charts are not produced at such a large size, the large plotter is used as follows. The computer is loaded with data tapes for four maps and programmed to drive the plotter so that all four are plotted consecutively (in different areas of the table, of course). The equipment may be left to operate without supervision overnight or over a weekend, thus being usefully employed for the maximum possible time. (An alternative technique is to use a smaller table with automatic wind-on of the plotting surface material.)

A computer cannot issue two commands to the plotter simultaneously; they have to be consecutive, although the time lapse is extremely brief. Hence, at any one moment the plotter can be commanded to move in the x or in the y direction, but not both. Thus a southwest to northeast line is plotted like a flight of stairs. However, the increments are generally so small (less than 0.01 mm in the best equipment) that the line appears to be smooth.

When plotting a string of digitized points, the plotter will travel by the shortest route (seen as a straight line) from one point to the next. If the points represent an irregular feature like a contour or a river and the spacing between the points is large, the plotted image will appear unnaturally angular with sharp changes of direction at each point. However, the computer can be programmed to compute a mathematical spline (a best-fit curve) through the points.

7.13 Plotting Modes

The (x,y) coordinates on the data tape are translated into movements of the x and y carriages on the plotter so that the plotting point arrives in the correct position. The label on the data tape records the nature of the feature digitized and hence the type of symbol by which it should be portrayed on the map. How does a plotter draw the required symbol?

In the simplest designs (such as desk-top graphplotters) it is only possible to mount a pencil or a pen. Various line colours can be drawn by changing the pen. Line width cannot be varied. Continuous lines and pecked lines can easily be drawn, with a limited number of variations such as cross-bars at intervals along a line. Simple point symbols such as circles and crosses can be drawn. To save time changing pens more advanced models have a set of pens (or even scribing tools) mounted in a rotatable turret; the required tool is brought into use automatically either by a code on the tape or by manual operation of a control panel switch. The drawing tool is held against the drawing surface under constant pressure (to give constant line width) by a magnetic field.

However, mechanical drawing is obviously not very suitable for the high-speed movement (up to 40 cm/s) of an automatic plotter. One possible advance would be to use a jet pen from which the ink is ejected without the pen actually touching the drawing surface.

Many of the problems in automatic plotting are solved by using a photographic method: a ray of light replaces the pen and the plotting surface has a photo-sensitive coating. The plotter is operated in a darkroom and the processed plot is a right-reading positive. Not only is friction between the plotting point and the surface eliminated, but there are many other advantages.

By varying the width of the light beam, the width of the plotted line is also varied. By interrupting the light in short periodic patterns, pecked lines in various styles can be drawn. An area can be covered with hatching. Parallel lines at various spacings can be drawn. The computer can be preprogrammed ('hard-wired') to draw *macros*, i.e. commonly used point symbols and alphanumerics in various styles; but it would be uneconomic to draw complex symbols (like hachures) in this way. Names can be output along a chosen azimuth starting from a given point. However, it should be noted that, for the reasons given in Section 7.9, programming the computer to plot names in the desired type size and style, position, orientation and spacing may consume more time than printing and positioning by the methods described in Sections 2.6.5 and 2.6.6. Another possibility is to keep symbols and alphanumerics in the form of negative transparencies and project them on to the plot.

7.14 Scanning Systems

An alternative technique for transferring the details of a graphic image into a taped record is similar to that used in television. The map is scanned by an electronic eye, line by line, each line being less than 0.1 mm wide. Over white paper the scan records 'no image' and over detail it records 'image'. On maps without elevation tints or area tints the point and line detail covers a very small proportion of the paper so that most of the *pixels* (units of record, spaced closer than 0.1 mm along each scanned line) will register 'no image'. Colour filters can be used to produce separate records of different colours in the original. The output from the taped record is, as in TV, generally displayed on a cathode ray tube. The technique has not found much popular acceptance in mapping, one major deficiency being that labels (feature codes) cannot be added during scanning.

7.15 Output on Microfilm

A technique known as COM (Computer Output on Microfilm) uses electronic equipment with no moving parts to plot from computer tape on to 35 mm microfilm or 105 mm microfiche, using a laser beam. If the film is diazo-sensitized it will develop as a negative. Positive enlargements can be made as in Section 8.6.1.

7.16 Electrostatic Plotters

The electrostatic printing process was explained in Section 5.17.5. The process is used in an electronic plotting system which 'draws' diagrams, plans, etc. much faster than is possible with a mechanically based plotter.

Data in digital (or vector) form must first be converted to raster form (see previous section), i.e. to a line-by-line scan format. A suitable computer programme will perform this operation. As each scan line of data is read by the plotter, the image pixels leave a charge on the paper while the non-image pixels leave no charge.

Paper is supplied from a roll up to 180 cm wide and passes through the plotter at up to 6 mm/s. The highest definition is about 8 pixels per millimetre so there can be over 14 000 pixels to each raster line. Fifty scan lines are exposed each second; after passing the toner applicator each charged pixel appears as a dot on the paper. Lines therefore plot as strings of dots.

Dot size is controlled so that each one fills the whole of one pixel space 0.125 mm square, thus a string of dots produces an unbroken line provided that the line is parallel to either the x or the y axis. In angled lines the 'staircase' effect is apparent to the unaided eye; for example a straight line on a grid bearing of 6° is made up of sections each 10 dots long running on bearing 0° with a sidestep of 1 dot between sections (see fig. 7.3).

This is not satisfactory for high-quality mapping; a definition of at least 20 dots/mm is probably the least value which would produce an acceptable result.

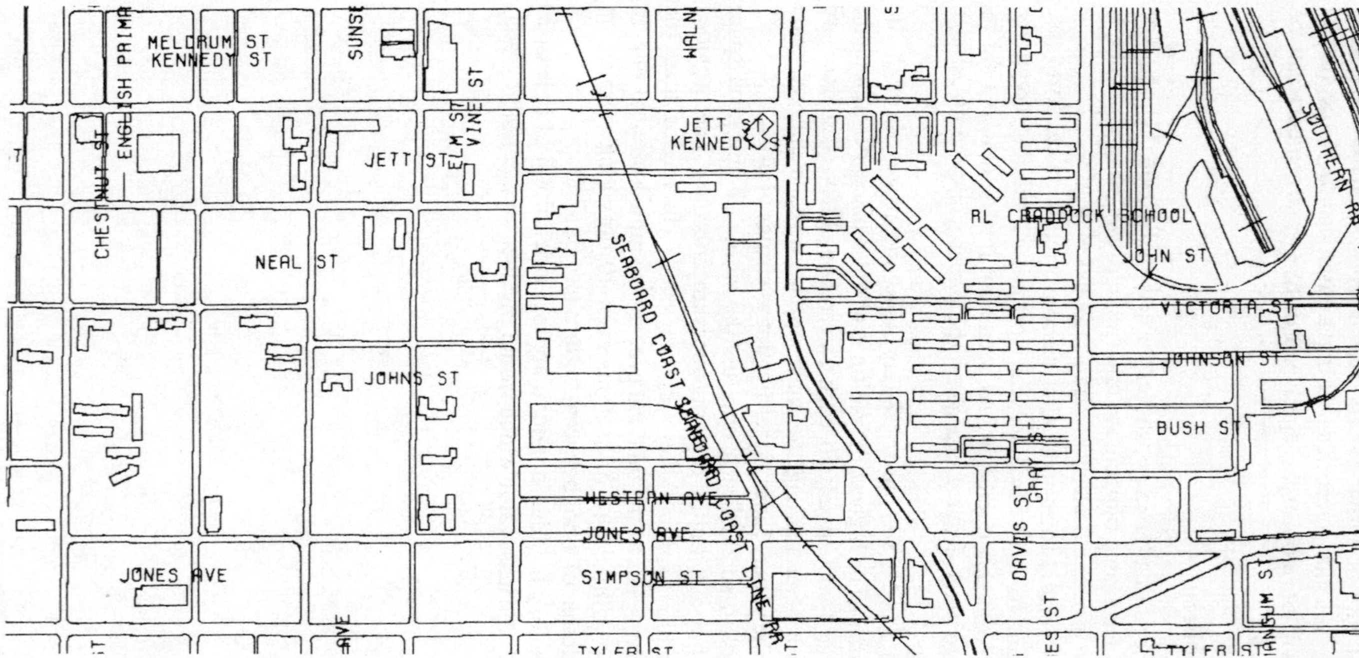

Fig. 7.3 (a) Map printed by electrostatic plotter

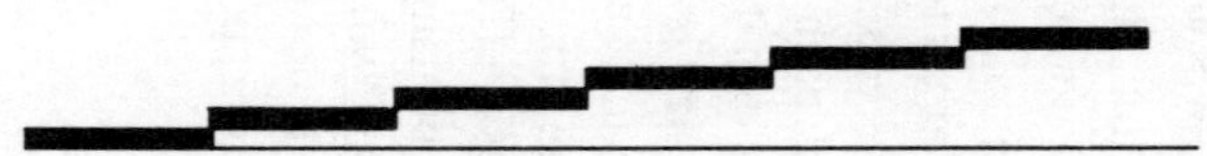

Fig. 7.3 (b) Enlargement of electrostatic plot

8

Map Records

8.1 Volume of Records

By the time a finished map starts rolling off the printing press, a great deal of material (paper, plastic, glass, metal) has probably been used along the production line; air camera film, diapositives, photogrammetric plots, reduction negatives, scribe sheets, opaque drawings, masks, reversals, combination negatives and positives, field completion sheets, name sheets, dyeline proofs, printing plate, etc.

Multiplying this total by the number (which may be hundreds or thousands) of different map sheets produced over a period of years must suggest that there is a problem of sorting, storing, indexing, and retrieval.

A large part of the graphic data of maps might be converted into digital computer data and stored on tapes, or the maps may be photographed on to microfilm (see below), both of which types of record are much less bulky, but there will inevitably be more or less graphic records to deal with.

8.2 Essential Records

The first step in imposing order on threatening chaos is a decision on what can be discarded; and the basic principle here is: if it can easily be reconstructed, let it go. The next decision is on what need only be stored temporarily and what must be kept as a permanent record, which in turn can be divided into archives and reference. Archives are preserved for posterity and should rarely be handled but reference material may be frequently needed and must be easy of access.

Air-photo film and prints are valuable until superseded by later photography of the same area; even then they still have an archival value for monitoring of and research into environmental change. Although positive prints are easier to examine, the negative film is the master document. Most of the material produced in subsequent stages of the production line can be discarded, except for the original field sheets of the lists of names and the final reversals from which the printing plates are made. The latter (plates) can be discarded unless there is a possibility of an early reprint.

Several record copies of the final printed map must be kept and at least one copy used as a revision record copy, on which reported errors and changes are to be noted until the next edition is prepared.

Other than the above copies, the printed maps are usually packeted in wrapping paper in bundles of 50 or 100 or 500 and kept in a bulk map store until issued for sale or other use.

It must be remembered that chemical images produced by any process (some are much more enduring than others) are liable to deterioriate during storage; they should not be included in any material destined for archives. Maps printed with good quality ink on good quality paper may retain their appearance for centuries if carefully stored. It is expected that xerographic copies should also keep well.

8.3 Storage Principles

The two basic essentials of good storage are

(*a*) safeguarding from damage or destruction
(*b*) accessibility for easy retrieval.

The greatest danger under (*a*) is fire. The store should be in a fireproof building and the stored material in fireproof cabinets. Adequate provision against fire will also give good protection against most other hazards such as earthquakes, rodents, insects, damp, dust, and riots.

For chemical images such as photofilm, control of temperature and humidity is essential. The storage space should be air conditioned.

In practice, the commonest cause of damage and deterioration of records is handling during removal and return. This can be considerably alleviated by efficient design and layout of storage space and equipment.

8.4 Storage Methods

The most primitive form of storage of unfolded map sheets is on open shelves, which is literally wide open to every possible hazard. A somewhat better and widely used method is in wooden or metal cabinets containing shallow horizontal drawers (not more than 5 cm deep). Clearly, one sheet per drawer would be uneconomical, but when there is more than one problems arise during return of a sheet to its proper place. Sheets get crumpled and may be pushed to the back of a drawer with the risk of falling to a lower level. To prevent this happening, a top cover may be fixed to the rear half of each drawer. Alternatively, the maps in a drawer may be placed in envelopes, five or ten to each envelope; such a package has sufficient rigidity, and one envelope may fairly easily be removed and later returned to its proper place. Another improvement is to have hinged fronts to the drawers so that maps or envelopes may be withdrawn without any lifting or bending of front edges. The highest drawer in a cabinet should not be above the shoulder level of the average user.

Horizontal drawers large enough to hold unfolded maps occupy a lot of floor space and need nearly as much again for access. Most of the disadvantages of a horizontal storage system are avoided by a vertical system, using a cabinet with a hinged opening top. There are several possible methods but since paper maps have no inherent rigidity to keep them flat when standing vertically on edge, it is essential that each sheet be individually suspended from its upper edge. The usual and most efficient way of doing this is to fix an adhesive strip of tough material (card, fibre, or plastic) along the full length of the north edge of the map. The strip has two or more pairs of holes in it (fig. 8.1). Horizontal bars pass through the holes, half the bars being fixed to the back of the cabinet, the other half to the front. The front is hinged at the bottom and can be opened a limited distance, at which the bars cease to overlap and selected maps can be lifted out.

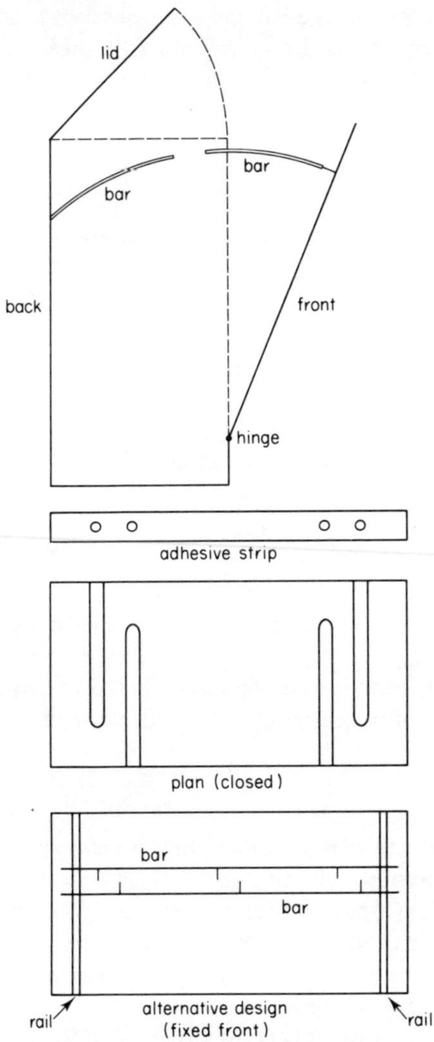

Fig. 8.1 Vertical map filing cabinets

An alternative pattern has a fixed front and pairs of horizontal bars running the width of the cabinet at the top, moving on rails at each side. Each pair of bars has horizontal prongs passing through the holes in the map mounting. The required map is located between one pair of bars and to remove it only that pair has to be separated.

The suspension strips are supplied with the cabinets and can be fixed to the maps in a special machine. The map reference number is written at the right-hand end of each strip for easy location. The system accommodates maps of different widths and lengths and it is possible (but not recommended) to suspend plastic envelopes with a loose map inside each.

Folded maps in printed covers can be stored like books, on shelves behind closed doors, preferably of glass or metal.

Printing plates are proof against most hazards and can be kept in open vertical racking. Reversals and drawings on plastic sheets thick enough to have some rigidity can also be stored in this way, or horizontally.

8.5 Working Space

> 'The actual arrangement of the whole map room must be left to the ingenuity of the individual planner.'
>
> (C. E. Le Gear (1956) *Maps in Libraries* Libr. of Congress, U.S.A.)

In all map storage areas (or map libraries) it is essential to have plenty of horizontal surfaces at a suitable level on which maps can be laid out for inspection.

Where horizontal filing drawers are used, the top of the cabinet may be suitable if it is not too high; otherwise tables may be needed. A common arrangement is to have the cabinets in the centre of the storeroom, back to back, the fronts facing windows under which inspection or working table tops are placed.

If the store is specially designed and built it is economical in ground area to have an upper level (either galleries or a complete floor) about two metres above the ground floor, on which further cabinets can be placed.

This is more efficient than having to use portable steps or ladders to reach high cabinets.

In designing a map store it must be remembered that thousands of stacked maps are heavy, and building strengths must be planned accordingly.

8.6 Microfilm

Microfilming has been in use for more than a century; its development was mainly due to requirements in time of war, when there is a high risk of destruction of records. Copies of files, ledgers, and other bulky documents can be made on reels of film, which have a much smaller volume, and can easily be transported to a safe place where they will be available for reference if the original documents are destroyed.

Camera film is made in various widths from 8 to 105 mm but the microfilm industry has tended to standardize on 35 mm. With suitable equipment, documents can be reduced 200 to 1 on film and the film later enlarged 150 times to give a readable product. The resolution and clarity of such an enlargement is not good enough for map detail, the practical limits for which are at present considered to be 30 to 1 reduction and × 13 enlargement; however, enlargements up to × 18.5 are frequently made. This is sufficient to produce copies from 35 mm film up to 650 mm wide. Within this limit it is possible to produce 40 × 40 cm copies of 1 : 2500 maps covering 1 × 1 km on the ground, or 56 × 56 cm copies of 1 : 50 000 maps covering ¼° squares.

8.6.1 Microfilm equipment

In a typical design the camera is mounted with lens axis vertical above a glass-topped copy table. Opaque originals on the table are lit from above, transparent ones from below. Controls are sited so that maps or other documents can be moved on to the table, photographed, and moved off at a rapid rate. A standard length of film is enough for 600 exposures.

Film can be developed in the normal way as a negative or by a reversal bleach technique to produce a positive. The film can be duplicated by a diazo process.

Film can be viewed in special enlarger projection equipment. Obviously positive film is preferable for this.

8.6.2 Microfilm retrieval

If the microfilm frames are serially numbered and an index made of the contents of the film, retrieval of a required frame is not difficult; exposure in an enlarger-printer will then supply a print at a desired scale. However it is found generally more convenient to mount a copy of each frame separately in a card. The frame occupies a hole in the card, hence the name 'Aperture card' (also known as a 'Copy-card'). Manufacturers can supply the cards with a piece of unexposed film already mounted in the aperture; the film is sensitized with silver halide or diazo according to whether it is required to produce positive from negative microfilm or from positive microfilm. Special equipment allows the cards to be exposed rapidly to successive frames of the film.

Cards are of a size suitable for filing in a standard card-index filing cabinet — 15 × 10 cm is a common size; 18 × 8 cm is also used. One drawer 20 cm deep will hold hundreds of cards; thus thousands of cards can be stored in a small fraction of the space required for cabinets of full-sized maps.

On each card is typed or printed the map reference number and any other useful information. Copies of the map can be made at the original scale in an enlarger-printer by a standard process (bromide, diazo, xerox, electrostatic).

If a monochrome map is produced of which no more than 200 copies are going to be needed it is cheaper to make them by this method than to go through the full lithographic process.

8.7 Map Library Index Systems

'There are probably as many individual map cataloging and classification systems in the world as there are map libraries.'

(J. WINEARLS (1967) *Proc. Assn. Canadian Map Libr.*, p. 27)

Most map producers accumulate not only the records of their own products but also printed maps from other producers, usually on an exchange basis. To get the maximum benefit from the resulting map library there must be an efficient indexing system and some working rules.

Every item received in the library should first be classified as archives, reference, or loan. Archives items are original material which cannot be replaced: air-photo film, field completion sheets, old editions of maps for which reproduction material is no longer available, etc. Access to and handling of archives material should be restricted to a minimum and only granted for good reasons. If there is regular demand it should be possible to provide duplicates which can be more freely accessible.

Reference material is material which has some special value, or is not easily replaceable. It may be freely consulted but generally should not be removed from the library.

Loan material most probably consists of maps of which the library either holds more than one copy or can easily get replacements. Loans may be internal (to other branches of the map production organization) or external. Since maps in flat sheet form are easily damaged, or distorted by rolling, and difficult to transport, maps for external loan should preferably be in folded form.

To retrieve a wanted map we must first find the drawer or cabinet in which it is stored. Map cabinets and other storage areas must be numbered on a system which allows the required number to be easily found. There are several international book library numbering systems (Library of Congress, Universal Decimal, Dewey Decimal, etc.) which can be applied to maps but these are devised for large libraries holding perhaps hundreds of thousands of maps. For a small library, a suitable system can easily be drafted locally.

Most demands on a map library first specify a location: a country, district, town, or other feature, then the purpose for which the map is required, e.g. the map showing the most detail, which implies the map on the largest available scale. The index should therefore be compiled in a way which will give the quickest and most complete answers to such demands.

Gazetteers are very useful reference books to support a map library index. If the demand starts, as it usually does, with a place name then the appropriate gazetteer should provide both the geographical coordinates and the political unit in which the place is located. Transferring the latitude and longitude to one or more index maps will reveal the serial number(s) of the map(s) covering the place.

Index diagrams (see Section 3.5) of map series showing sheet numbers and names and graticule or grid values of sheet edges are essential in a map library. They not only show the ground coverage of available map sheets but can also be used to indicate where maps are in preparation or proposed.

The most flexible form of conventional index is a card index. Details of each map or map series are noted on a card starting with continent or country for small-scale maps and national series, and going on to town or other location (e.g. national park) for larger-scale maps. The other necessary details to be entered are: scale, whether single sheet or series, edition number and date (for single sheets), number of sheets (for series), whether topographic or thematic, publisher, contour vertical interval, projection, language, etc.

The above data will normally suffice for most purposes but for some indexes any or all of the following may also be included:

Latitude and longitude of sheet centre
Dimensions of paper and of mapped area
Datums and grid
Dates of air photo, field completion, revision, etc.
Printer, holder of copyright
Availability on paper, plastic, microfilm, computer tape, etc.

For a series, a single card is usually considered sufficient. There is unlikely to be sufficient space to give details of sheet names and numbers and sheet lines: the reader is directed to refer to a sheet index diagram for further information. This may be supplemented by a list showing latest edition number and date for each sheet.

8.8 Computerized Catalogues

Large book libraries (such as Library of Congress in Washington) already have catalogues in the form of computer data banks and the system is being applied to large map collections. Map librarians should become familiar with such terms as MARC (machine readable cataloging), DILS (data integrated library system), and CAFS (content addressable file store). Data may be recorded on magnetic tape or disc, the latter giving quicker retrieval. The data elements are much the same as those entered in a card index (see above). An enquirer might ask for a large-scale contoured map of, say, Timbuktu; these details are fed into the computer which will search the data bank and display on a CRT the details of any map held in the library which meets this demand. Output may also be by machine print-out. Obviously large libraries can have wired connections between their data stores or can supply each other with duplicate tapes, so that the system could operate on a world-wide basis between willing partners.

> 'Any map library planning to automate its cataloging should use the MARC format.' (LARSGAARD, p. 147)

8.9 Printed Catalogues

A well-organized map store/library supported by an efficient index is excellent for internal use by map producers and their visitors; but information about maps available for sale should also be available in easily portable form for distribution to anyone interested, e.g. reference libraries all over the world, bookshops, tourists, and others having business which requires topographic information.

The form and content of such a catalogue are determined by financial considerations. Ideally it should be issued free on demand as a form of sales publicity. This may imply an economical abridged format (the cheapest being a single sheet printed both sides and produced in folded form). The cost may be recovered from increased sales of maps. It can be argued that there should be a small nominal charge to prevent reckless waste, as is often the fate of free handouts.

It may be possible to cover part of the cost of a catalogue by selling advertising space in it. To interest advertisers, an attractive appearance and wide distribution are necessary. Putting a sale price, however low, on a catalogue will restrict distribution, publicity, and consequent increased sale of maps.

Catalogues are usually published in booklet form. The contents should include:

(*a*) an introduction listing addresses where the maps can be seen and purchased, and stating how to order by post.

(*b*) a description of the maps available. Details given should include: scale, whether sheet or series (and number of sheets), area covered, sheet size, number of colours, contour interval, topographical or thematic, date(s) of publication, and whether lithographic or other form of copy.

A specimen extract to illustrate each class of map is desirable but rarely included.

The descriptions should be arranged in logical groups, e.g. all small- and medium-scale topographical maps, all town maps, charts, special and thematic maps, etc.

(*c*) index diagrams for series maps and to show location of special maps.

(*d*) price list.

(*e*) information on other available related material such as air photos, microfilm cards, diagrams, computer tapes, etc.

8.10 Costing and Pricing

Commercial cartographers must recover costs of production from sales; their products are subject to normal trading principles, i.e. they must supply a demand, they should have sales-appeal, their price should be low enough to sell easily but high enough to produce a profit, and so on.

Production of basic mapping is quite different. Commercial costing of ground and air survey and subsequent operations would put the sale price of ordinary topographical maps so high there would be hardly any market. It is possible to sell enough copies of a map of a popular tourist area to recover the costs of that particular sheet but the majority of sheets of a national map series do not cover such areas and cannot be profitable.

Basic mapping is an essential service to the community. Any physical development on or below or above land or water cannot be properly planned or controlled without suitable maps and there are a great many other communal or official activities which equally need maps. Basic mapping should undoubtedly be regarded as a proper charge on public funds. Map sale (and other related service) prices must be determined by factors other than raising of revenue; it may, however, be worth while consulting a professional marketing research organization to find a sale price which yields maximum returns.

Bibliography

American Cartographer periodical 1974-) A.C.S.M., Washington, DC, U.S.A.

Bomford, A. G. (1979). *The Role of a National Mapping Organisation*, Survey Review Vol. XXV no. 195, pp. 195–210.

Canadian Cartographer (periodical 1964-) Toronto.

Cartographic Journal (periodical 1964-) Brit. Carto. Soc., England.

Cartography (periodical 1954-) Australian Inst. Carto. Canberra.

Glossary of Technical Terms in Cartography (1966). British National Committee for Geography, Royal Society, London.

International Year Book of Cartography (1961-). Kirschbaum, Bonn.

Jackson, J. E. (1981). *Sphere, Spheroid and Projections for Surveyors*, Granada, London.

Keates, J. S. (1973). *Cartographic Design and Production*, Longman, London.

Koeman, C. (ed.) (1980). *I.C.A. Basic Manual of Cartography*, Wiley, London.

Maling, D. H. (1973). *Coordinate Systems and Map Projections*, G. Philip, London.

Manual of Field Completion Survey. Surveys and Mapping Branch, Ministry of Resources, Ottawa.

Military Engineering Vol. **XIII** (1971) Part 12, *Cartography*, Ministry of Defence, London.

Military Engineering Vol **XIII** (1976) Part 13, *Map Reproduction*, Ministry of Defence, London.

Larsgaard, M (1978). *Map Librarianship*, Libraries Unlimited, Littleton, Col., U.S.A.

Nichols, H. (1982). *Map Librarianship, Edn 2*, Bingley, London.

Phillips, R. J., Noyes, E., and Audley, R. J. (1978). *Names on Maps, Cartographic Journal* Vol **15**, No. 2, pp. 72–77.

Richardus, P., and Adler, R. K. (1975). *Map Projections*. North Holland.

Robinson, A., Sale, R., and Morrison, J. (1978). *Elements of Cartography*, Wiley, London.

Tissot, A. (1881). *Memoire sur la Répresentation des Surfaces et les Projections des Cartes Géographiques*. Paris.

Tyrell, A. (1972). *Basis of Reprography*, Focal Press, London.

World Cartography (periodical 1951-). United Nations, New York.

Appendix

Training for Map Production

A map production organization generally has a staff 'pyramid' like any other large business establishment. Those members of staff with special skills relating to some aspect of map production may be broadly classified as professional, technologist, or technician.

Technicians usually specialize in one stage only of the production line such as drafting or printing. Training may be entirely on-the-job within the organization or preferably supplemented by evening classes, day-release, or sandwich courses at a technical college. Successful completion of such courses leads to the award of some recognized certificate or diploma.

Technologists (engineers in America) are also specialists who study theory and practice of a particular technique such as photogrammetry, automation, or reproduction in greater depth; within their particular field they can offer advice and expertise to the professional staff. For some technologies there are specialized training institutions; some polytechnics and colleges of advanced technology may offer suitable courses.

Professional staff should be familiar with every aspect of map production, although they cannot hope to master all the techniques in depth or become expert practitioners in more than one or two. Acceptable qualifications are a relevant university degree and/or membership of an appropriate professional body. However Cartography and Map Production are not yet recognized as full first degree subjects (except in Russia) nor is there any professional society conducting examinations limited solely to either.

Therefore the relevant degree or professional qualification must be one with a high cartographic content. Degrees in Land Surveying, Topographical Science or Environmental Science are now awarded at universities in many parts of the world and generally appear to be the most suitable. Some geography degrees have a surveying and mapping option which may make them acceptable.

The International Institute for Earth Sciences (ITC) at Enschede in the Netherlands offers several suitable courses (conducted in the English language) to international students of all ages. There are one year courses for cartographic technicians and engineers and a one year postgraduate diploma in Cartography for holders of a relevant degree. Holders of the diploma may continue another year for the M.Sc. (Cartography) degree.

Index